T0313906

*Time Series Analysis in the Social Sciences*

# Time Series Analysis in the Social Sciences

## THE FUNDAMENTALS

*Youseop Shin*

UNIVERSITY OF CALIFORNIA PRESS

University of California Press, one of the most distinguished university presses in the United States, enriches lives around the world by advancing scholarship in the humanities, social sciences, and natural sciences. Its activities are supported by the UC Press Foundation and by philanthropic contributions from individuals and institutions. For more information, visit www.ucpress.edu.

University of California Press
Oakland, California

Library of Congress Cataloging-in-Publication Data

Names: Shin, Youseop, 1964- author.
Title: Time series analysis in the social sciences : the fundamentals / Youseop Shin.
Description: Oakland, California : University of California Press, [2017] | Includes bibliographical references and index.
Identifiers: LCCN 2016022158 (print) | LCCN 2016023292 (ebook) | ISBN 9780520293168 (cloth : alk. paper) | ISBN 9780520293175 (pbk. : alk. paper) | ISBN 9780520966383 (ebook)
Subjects: LCSH: Time-series analysis. | Social sciences—Statistical methods.
Classification: LCC QA280 .S55 2017 (print) | LCC QA280 (ebook) | DDC 519.5/5—dc23
LC record available at https://lccn.loc.gov/2016022158

26   25   24   23   22   21   20   19   18   17
10   9   8   7   6   5   4   3   2   1

*To my parents, Dongjoon and Hangil,*
*my wife, Jungwon,*
*and my son, Lucky Boy, Jaeho*

CONTENTS

In the social sciences, it generally takes longer to collect temporal data than to collect cross-sectional data. In addition, it is generally hard to obtain observations for as many time points as are needed for good time series analysis. For these reasons, time series analysis is employed less frequently than cross-sectional analysis. Time series analysis, however, can serve special purposes.

For example, we may identify and explain a systematic temporal pattern of a variable, even when the variable does not appear to change significantly over time. We can explain and predict the dependent variable from the observations of its past behavior, instead of relying on a limited set of independent variables. Thereby, any independent variables that may influence the dependent variable are taken into account, and we can avoid threats to external validity of our explanations of the dependent variable. In multiple time series analysis, we can reflect any excluded important independent variables in a lagged dependent variable on the right-hand side of the equation. We can estimate how long an estimated effect of an independent variable on the dependent variable persists. By comparing two trends before and after a program is implemented, we can test whether the program brings about the expected effect.

To employ an appropriate time series analysis when necessary, social scientists and professionals, such as policymakers, campaign strategists, securities analysts, and realtors, should first understand what time series analysis is, and what we can do with it, and how. There are many books on time series analysis. However, there are no books that carefully and gently explain time series analysis for social scientists and professionals. Most books on time series analysis contain sophisticated mathematical materials, such as matrixes and integral calculus, which most social scientists and professionals do not have to understand for their research and practice.

The purpose of this book is to provide practical, easy-to-understand guidelines for time series analysis that are suitable for social scientists and professionals. I intend to make this book a primer for social scientists and professionals which helps them learn and utilize time series analysis without having to understand extraordinary mathematical materials, which they are not likely to use when they employ time series analysis as part of their research and practice. Knowledge of regression analysis, that is, how to estimate a slope and what are the properties of residuals, can be helpful in understanding this book. Knowledge of basic algebra can also be helpful in understanding this book.

This book does not include everything about time series analysis. Instead, it focuses on the most fundamental elements that social scientists and professionals need to understand to employ time series analysis as part of their research and practice. In this book, I explain univariate time series analysis step by step, from the preliminary visual analysis, through the modeling of seasonality, trends, and residuals, to the prediction and the evaluation of estimated models. Then I explain how to conduct multiple time series analysis and interrupted time series analysis.

At each step, I explain time series analysis, not statistical programs. I provide general explanations about how to conduct statistical analysis, not focusing on a particular statistical program, except in a few cases in which it is necessary to caution readers about specific procedures of a particular program. Readers are expected to be able to calculate statistics with a calculator and statistical programs' compute procedure, if any of these statistics are not reported by the reader's statistical program. For readers who need to learn how to conduct time series analysis with their statistical program, I list websites for EViews, MATLAB, Minitab, R, S+, SAS, SPSS, Stata, and Statgraphics in appendix 1. These websites provide detailed explanations of how to conduct time series analysis with these programs.

At the end of each step, I provide an actual analysis of monthly rates of violent crimes (murder, forcible rape, robbery, and aggravated assault). The data were compiled from *Uniform Crime Reports: Crime in the United States* (Washington, DC: Department of Justice, Federal Bureau of Investigation). For univariate and multiple time series analysis (chapters 2–6), I model the monthly violent crime rates from 1983 to 1992 and use the monthly crime rates of 1993 to evaluate the estimated model. For the interrupted time series analysis in chapter 7, I analyze monthly violent crime rates from 1985 to 2004, ten years before and after the implementation of a tougher crime-control policy in 1994.

By employing the same example across chapters, I intend to help readers understand the procedure of time series analysis coherently and synthetically. Readers are expected to move back and forth to compare discussions across chapters. For example, readers can compare a case where we treat residuals as noise with another case where we treat residuals as conveying important information on the relationship between the dependent variable and an independent variable. These comparisons are more intuitive when the same example is used than when disconnected multiple examples are used. Readers can directly see the differences when they compare the same example.

Examples that can draw attention will vary from discipline to discipline and from reader to reader, and no one book can successfully cover all of them. I provide the monthly violent crime rates from 1983 to 2004 (appendix 2) and a list of webpages/data resources from which readers can download social science time series data for their own use (appendix 3).

Chapter 1 explains how time series analysis has been applied in the social sciences. Chapter 2 defines important concepts and explains the structure of time series data. Then it explains the univariate time series modeling procedure, such as how to visually inspect a time series; how to transform an original time series when its variance is not constant; how to estimate seasonal patterns and trends; how to obtain residuals; how to estimate the systematic pattern of residuals; and how to test the randomness of residuals. Chapter 3 explains diagnostics. Several properties of residuals should be satisfied, if the fitted model is appropriate. Residuals should be a realization of a white or independent and identically distributed (IID) sequence. They should have zero as a mean. Their variance, $\sigma^2$, should be constant. They should be normally distributed as well. This chapter explains how to test these points. Chapter 4 explains how to forecast future values based on the estimated time series model and how to evaluate the accuracy of the estimated model. Chapter 5 explains how to make trends stand out more clearly by reducing residual fluctuations in a time series, focusing on two widely employed techniques, exponential smoothing and moving average smoothing. Chapter 6 applies the above explanations to time series analysis with two or more time series variables, such as cross correlation and bivariate or multiple time series analysis. In multiple time series analysis, the dependent variable is the monthly violent crime rate, and the independent variables are unemployment rate and inflation. This chapter discusses several topics related to the robustness of estimated models, such as how to prewhiten a time series, how to deal with autoregressive residuals, and how to discern changes in the dependent variable caused by independent variables from its simple continuity. In addition, this chapter discusses the concepts of cointegration and long-memory effect and related topics such as error-correction models and autoregressive distributive lag models. Chapter 7 explains interrupted time series analysis. This chapter includes the impact analysis of the Three Strikes and You're Out law, with October 1994 (when Public Law 103–322 was enacted) as the intervention point.

I owe thanks to the many people who helped me complete this book. These people include my professors who taught me statistics and research methods—Aage Clausen, Paul Diel, Dan Durning, Robert Grafstein, Timothy Green, Paul Gurian, Patrick Homble, Snehalata Huzurbazar, Edward Kellough, Brad Lockerbie, Nancy Lyons, Ashim Mallik, Lynne Seymour, and Herbert Weisberg—and my students, and four anonymous reviewers. I worked on this book in Room 748, Barrows Hall, while at the Department of Political Science, University of California, Berkeley, as a Fulbright scholar. I thank the department and Taeku Lee and Eric Schickler (chairs), Lowell Dittmer, Robert Van Houweling, Hongyoung Lee, and other professors for their help and encouragement. I thank the Fulbright Program and Jai-Ok Shim, Marilyn Herand, and people who work for the program for their assistance. I am grateful for the guidance and encouragement of the University of California Press editor, Maura Roessner. I also thank Francisco Reinking, Roy Sablosky, Jack Young, Sabrina Robleh, Chris Sosa Loomis, and the people at UC Press for their support in producing this book. I want to express the greatest thanks to my family and friends. Their love, encouragement, and faith in me have made it possible for me to overcome many difficulties that I could not have overcome otherwise.

# ONE

# Time Series Analysis in the Social Sciences

IN THE SOCIAL SCIENCES, data are usually collected across space, that is, across countries, cities, and so on. Sometimes, however, data are collected across time through repeated regular temporal observations on a single unit of analysis. With the data that are collected in this way and that are entered in chronological order, we explore the history of a variable to identify and explain temporal patterns or regularities in the variable. We also explore the relationships of a variable with other variables to identify the causes of the temporal pattern of the variable.

Time series analysis is not as frequently employed in the social sciences as regression analysis of cross-sectional data. However, this is not because time series analysis is less useful than regression analysis but because time series data are less common than cross-sectional data. It is the characteristics of the data at hand, not the usefulness of statistical techniques, which we consider to select between time series analysis and regression analysis.

When we deal with time series data, time series analysis can be more useful than ordinary least squares (OLS) regression analysis. Employing OLS regression analysis, we cannot appropriately model a time series, specifically its systematic fluctuations, such as seasonality and systematically patterned residuals (see chapter 2). As a result, the standard errors of regression coefficients are likely to be biased, and independent variables may appear to be statistically more significant or less significant than they actually are.

Time series analysis can be employed in several ways in the social sciences. The most basic application is the visual inspection of a long-term behavior (trend) of a time series (see chapter 2). For example, in order to survey the extent of partisan change in the southern region of the United States, Stanley (1988) visually inspected the percentages of Republicans, Democrats, and

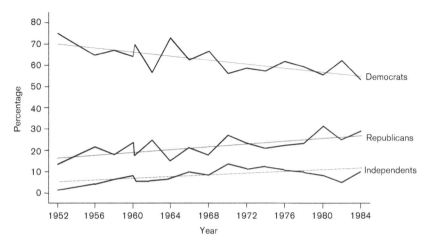

FIGURE 1. Party identification in the South, 1952–1984.

SOURCE: Stanley (1988), Figure 1, p. 65. Reproduced with permission of the University of Chicago Press.

independents from 1952 to 1984. As can be seen in figure 1, visual inspection is enough to show that both realignment and dealignment characterized southern partisan changes and that it was the Democratic Party that suffered the most from the change.

When we estimate trends, time series analysis is bivariate OLS regression analysis where the independent variable is Time with a regular interval. This time series analysis is called *univariate time series analysis* (see chapter 2). The trend in time series analysis is the slope of Time in bivariate OLS regression analysis. For example, Cox and McCubbins (1991) regressed the percentage of times individual legislators voted with their party leaders from the 73rd to the 96th Congress on Time. They showed that party voting significantly declined only for the Republicans (figure 2).

In many cases, a time series contains systematic short-term fluctuations other than a long-term trend. That is, observed values increase for a certain period and decrease for another period, rather than randomly fluctuating over the fitted linear line. These systematic patterns in time series variables should be removed to examine accurately the relationship between them. When systematic patterns are present in two time series variables, the correlation between the two can simply be a product of the systematic patterns (see chapters 2 and 6).

For example, Norpoth and Yantek's (1983) study of the lagged effect of economic conditions on presidential popularity raised a question about Mueller

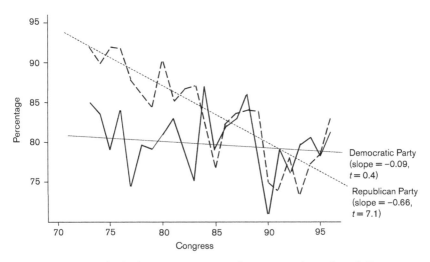

FIGURE 2. Average leadership support scores on the party agenda, 73rd–96th Congress.
SOURCE: Adapted from Cox and McCubbins (1991), Figures 1 and 2, pp. 557–558. Reproduced with permission of John Wiley & Sons Inc.

(1970, 1973), Kramer (1971), and Kernell (1978). Their estimates of economic effects, according to Norpoth and Yantek, are vulnerable to serial correlation within the independent variables, the monthly observations of unemployment or inflation. Norpoth and Yantek identified stochastic processes (ARIMA, explained in chapter 2) for the inflation series and for the unemployment series. They removed the estimated stochastic processes from the observed series. Since the inflation series and the unemployment series were no longer serially correlated, the relationship between inflation (or unemployment) and presidential popularity could not be an artifact of autocorrelation of the inflation series or of the unemployment series.[1] Norpoth and Yantek found that past values of the inflation and unemployment series did not significantly influence current approval ratings of presidents with any particular lag structure. This finding is not in accord with the conventional wisdom that the economy matters for presidential popularity and also for presidential electoral outcomes. Studies of presidential elections (Key 1966; Lewis-Beck 1988; Lockerbie 1992) present evidence that the national economy and the evaluation of a president's handling of the nation's economy do matter for the public support for the president. Norpoth and Yantek are reluctant to conclude that inflation and unemployment do not influence presidential popularity. They discuss some problems raised by the removal of the estimated stochastic processes from the observed series. Nonetheless, Norpoth and Yantek show that it is possible to lead to a very different finding when we ignore serial correlation in a time series versus when we remove it from the series.

Autocorrelation among residuals is a serious violation of a vital assumption concerning the error term in regression analysis (Achen 1982; Berry and Feldman 1985; Lewis-Beck 1980), and it is very likely to be present when we collect observations across time. With the serially correlated residuals, the least-squares estimates are still unbiased but may not be the best, with the minimum variance. Also, the significance tests and the confidence intervals for regression coefficients may be invalid. With time series analysis, we can directly check and estimate a systematic pattern that remains after we fitted a trend line to a time series (see chapter 3). If we are concerned only about the trend estimation, we can remove the systematic pattern from residuals before we fit a trend line to a time series by smoothing the time series (see chapter 5). In multiple time series analysis, we can deal with autocorrelated residuals in several different ways (see chapter 6). For example, we can estimate and then eliminate systematic patterns from each time series before we conduct multiple time series analysis. Alternatively, we can estimate a multiple regression model with autoregressive processes by adjusting regression coefficients according to estimated autocorrelation coefficients.

Once we estimate an autoregressive process of a time series, we can utilize the autoregressive process to determine how long the time series's time-dependent behavior or its impact on the dependent variable will persist (see chapter 6). For example, comparing the autoregressive parameters of two independent variables, Mackuen, Erikson, and Stimson (1989) show that the impact of consumer sentiment on aggregate-level party identification lasts longer than that of presidential approval, although the immediate impact of the former is smaller than that of the latter. As a result, the total impact of consumer sentiment on party identification is greater than that of presidential approval.

Time series analysis has perhaps been employed most frequently to forecast future outcomes, for example of presidential elections (see e.g. Lockerbie 2004, 2008; Norpoth and Yantek 1983; Norpoth 1995; Rosenstone 1983). In interrupted time series analysis, forecasted values are used as *counterfactuals* that represent a time series that we would have observed had there not been an intervention, such as the implementation of a policy (see chapter 7). We compare forecasted values with observed values to determine whether an intervention has the intended impact (see e.g. McCleary and Riggs 1982; Mohr 1992).

We can forecast future values by utilizing the information of past observations of a time series itself (see chapter 4) or by referring to the estimated relationship of the dependent time series variable with other time series variables (see chapter 6). In the latter case, we can forecast with greater accuracy, as the model's coefficient of determination is larger. However, we cannot know in advance what will be the exact values of predictor variables, even in the near future. Without the information of predictor variables in the future, forecasting with the estimated relationship of the dependent time series variable to other time series variables is only making a guess. In this case, forecasting with the estimated multiple time series analysis model could be worse than forecasting with the information of the past behavior of the dependent variable itself.

In addition, our multiple time series analysis model is generally not exhaustive. Future values forecasted by referring to the behavior of a time series itself may be more accurate than those forecasted by referring to the estimated relationship of the dependent time series variable with a few select independent variables. The behavior of the dependent time series variable will reflect influences from all factors that should be included in a multiple time series model.

However, our explanation of the forecast will be limited when we forecast future values by utilizing the information of past observations of a time series itself: we can provide explanations about our forecasts only in terms of the behavior of the time series but not in terms of factors that cause changes in the time series. Presidential-election outcomes, for example, can be influenced by various factors, such as the electorate's positions on salient issues, the state of the economy, characteristics of the candidates, presidential popularity, and evaluation of presidential job performance. With multiple time series analysis, we can provide explanations of our forecasts in terms of the relationships between these factors and presidential-election outcomes.

When we forecast future values by referring to the behavior of a time series itself, systematic patterns in residuals are important components of the time-dependent behavior of the time series. Without estimating such systematic patterns, our model is not complete and we cannot accurately forecast future values. For example, when we forecast future values with a time series with no discernible trend, as in figure 3, the trend line that is estimated with OLS regression analysis does not convey meaningful information. Even in this case, we may still identify underlying systematic patterns in residuals. Different systematic patterns will require different interpretations of the behavior of a time series and lead to different forecasted values. Figure 3, for example, can be interpreted in two different ways (Norpoth 1995). First, if there is no

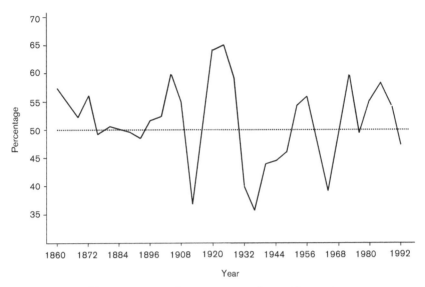

FIGURE 3. Republican percentage of major-party vote for president, 1860–1992.
SOURCE: Norpoth (1995), Figure 1, p. 202. Reproduced with permission of Cambridge University Press.

systematic pattern, that is if the time series consists of independent and identically distributed noise,[2] we can say that the chance of winning the election is equally divided between the two political parties. To put it differently, there is perfect competition between the two parties. Second, if some type of dependency exists among the observations, that is if there is an autoregressive pattern in the series, we can say that one party tends to have advantage over the other for a certain period of time, and then the relationship is reversed for another period of time. By identifying an autoregressive model, we can estimate when the reversal of electoral advantage tends to occur.

In univariate time series analysis, we can estimate seasonality and systematic patterns of residuals and thereby improve the accuracy of our model and forecast. However, short-term regularities in residuals may be artifacts of factors that are not considered in univariate time series analysis. To take this possibility into consideration, we may employ multiple time series analysis that includes these factors (see chapter 6). The public approval of the president, for example, tends to increase for a certain period of time and to decrease for another period of time. However, this may be not because the public approval of the president is serially correlated but because economic conditions, such as unemployment and inflation, are serially correlated. If those factors show random fluctuations, presidential popularity may also randomly change.

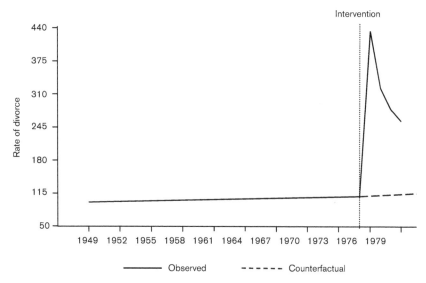

FIGURE 4. The 1975 Family Law Act and divorce rates in Australia.
SOURCE: Adapted from McCleary and Riggs (1982), Figure 1, p. 11. Reproduced with permission of John Wiley & Sons Inc.

Time series analysis is frequently employed in the impact analysis of a policy. Time series analysis used for this purpose is called *interrupted time series analysis*. The basic idea of interrupted time series analysis is that, if a program has an intended impact on the behavior of a target, the pre- and post-intervention segments of the temporal observations of the behavior should show significantly different levels and/or trends (Cook and Campbell 1979; McDowall et al. 1980; Mohr 1992). To put it differently, if a program has any significant impact, such an impact will be identified, as in figure 4, by the difference between the observations after the intervention point and the *counterfactual*, the projection of the correctly-modeled-before series into the post-intervention period (see chapter 7).

We can also test the impact of one or more factors on the dependent variable by comparing the observed values of the dependent variable with the values forecasted by the factors. For example, Mackuen et al. (1989) show that the observed macropartisanship—the percentage of Democrats divided by the percentage of Democrats plus Republicans—matches well the one predicted by presidential approval and consumer sentiment, meaning that the macropartisanship has been shaped by these short-term forces. In order to show that party identification is more stable than Mackuen et al. argue, Green,

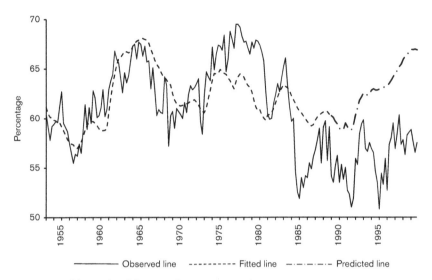

FIGURE 5.  Observed and forecasted macropartisanship.
SOURCE: Adapted from Green et al. (2001), Figure 19-3, p. 362. Reproduced with permission of CQ Press.

Palmquist, and Schickler (2001) compared the observed macropartisanship with the forecasted one (figure 5). The forecast was based on a model that includes presidential approval, along with the lagged dependent variable and the systematic pattern of residuals, thereby controlling out the time-dependent behavior of the dependent variable. Green et al. show that when the estimated model is applied to new data points, macropartisanship is not well forecasted with presidential approval and thereby conclude that macropartisanship is stable to a greater extent than suggested by Mackuen et al.

Sometimes a variable is observed for a number of cross-sections over a time span. Combining cross-sectional data and time series data is useful when the number of available time points is too small and/or the sample of cross-sections is small in size (Sayrs 1989). The characteristics of data may also require us to employ a pooled cross-sectional analysis. Markus (1988), for example, includes in his model one individual-level variable (personal financial change) and one aggregate-level variable (the annual rate of change in real disposable personal income per capita). The latter can be cross-sectionally observed only in a comparative study of multiple nations. Markus, however, analyzes the presidential vote in only one nation, the United States. Thus, in order to estimate the impact of the aggregate-level variable on the presidential vote, he decides to cover a long period, from 1956 to 1984.

Time series analysis will continue to be a major interest in various social science disciplines. The purpose of this book is to provide easy-to-understand, practical guidelines for time series analysis to social scientists who want to understand studies, such as those cited above, that utilized time series analysis techniques or who want to conduct their own research and practice.

# TWO

# Modeling

## (1) PRELIMINARY DEFINITION

A TIME SERIES IS COLLECTED across time through repeated regular temporal (daily, weekly, monthly, quarterly, or yearly) observations on a single unit of analysis and entered in chronological order. It is different from cross-sectional data, which are collected across space and/or across groups at the same time point. In cross-sectional analysis, the level of observation should be fixed and constant. Otherwise, observations cannot be compared with each other. When analyzing crime rates across countries, for instance, we should collect crime rates per a predetermined size of population. In time series analysis, in addition to the level of observation, the temporal interval of observation should be fixed and constant. Otherwise, any observed temporal pattern can hardly be meaningful. In cross-sectional data, the number of cases is the number of individuals or groups included in the data. In time series data, the number of cases is the number of repeated regular temporal observations.

A *lead-lag relationship* exists between observations in a time series. An observation leads (lags) another if the former occurs earlier (later) than the latter. A lead-lag relationship can also exist between two time series variables. When a time series, $\{Y_t\}$, is significantly correlated with another time series that is shifted $k$ lags forward ($k$ is a fixed certain integer greater than zero), $\{X_{t+k}\}$, we say that $\{Y_t\}$ leads $\{X_t\}$. When $\{Y_t\}$ is significantly correlated with another time series that is shifted $k$ lags backwards, $\{X_{t-k}\}$, we say that $\{Y_t\}$ lags $\{X_t\}$. This lead-lag relationship is not necessarily a causal relationship. However, for a variable to influence another, the former needs to lead the latter.

A time series consists of a trend, seasonality, and systematically patterned residuals. The purpose of univariate time series analysis is to estimate these three major components. When these components are removed from a time series, only white noise should be left. *White noise* is a sequence of uncorrelated random values that has a zero mean and a constant variance.

A *trend* in univariate time series analysis is the slope in a bivariate ordinary least squares (OLS) regression model with a time series as the dependent variable. The independent variable is Time, whose values are fixed and constant temporal intervals at which the time series was measured. A line is fitted to a time series in a way that minimizes the sum of squared errors, that is the sum of squared distances between the observed values and the values that are predicted by the estimated OLS regression model. Trends measure the average amount by which this line inclines or declines per unit change of the independent variable, that is, the average change of the dependent variable associated with the level of a regular time interval at which the dependent variable was measured. This type of trend is predictable, and therefore it is called a *deterministic trend*.

A time series often contains cyclical regular fluctuations with a period of less than one year, such as day, week, month, or quarter. For instance, soft-drink sales are likely to have a seasonal pattern, with a peak in summer and trough in winter. These cyclical regular fluctuations are called *seasonality*. Whenever seasonality is present in a time series, it should be estimated and removed from the time series for more accurate trend estimation.

In OLS regression analysis of cross-sectional data, we are usually unconcerned about *residuals*, which are left unexplained by the estimated model. We acknowledge their presence, but we rarely attempt to analyze them. In univariate time series analysis, residuals are often of major concern. We estimate systematic patterns of residuals, such as an *autoregressive* process of order $p$, $AR(p)$, and/or a *moving average* process of order $q$, $MA(q)$, after we estimate and eliminate the trend and seasonality from a time series (see section 4). This pattern is called a *stochastic trend*, as compared with a deterministic trend, because its values are not predictable in the long term.

Sometimes, we may be interested only in systematic patterns of residuals. In this case, we skip the estimation of trend and seasonality and just remove them from the time series. The process of removing trend and seasonality from a time series without estimating them is called *differencing* (see section 4). For example, to remove a linear trend from a time series, we difference the series at lag 1. Seasonality can be removed by differencing a time series at lag 12, if the time series consists of monthly observations, or at lag 4, if the time series consists of quarterly observations, and so on.

A time series, $\{X_t\}$, is (*weakly*) *stationary* if it has the same mean and variance as those of any time-shifted series, $\{X_{t+k}\}$, and if the covariance between $\{X_t\}$ and $\{X_{t+k}\}$ is constant over time for any $k$ that is a fixed certain integer greater than zero.[1] A time series, $\{X_t\}$, is *strictly stationary* if not only its mean, variance, and covariance but also the joint distributions between $\{X_t\}$ and $\{X_{t+k}\}$ are invariant for all $t$ and $k$ ($1 < t < n$; $k >$ o). An example of a weakly stationary time series is an $AR(p)$ process with an AR coefficient smaller than 1 (see chapter 6, section 3B). White noise is a stationary time series with mean zero, a constant variance, and a zero covariance between $\{X_t\}$ and $\{X_{t+k}\}$ ($k > 0$). An example of a strictly stationary time series is independent and identically distributed (IID) noise. IID noise is a sequence of uncorrelated random values with a zero mean and a zero covariance between $\{X_t\}$ and $\{X_{t+k}\}$ for all $t$ and $k$ ($k > 0$). In addition, the value and variance at a time point are not dependent on previous values and variance(s).[2]

Under the normal distribution, zero correlation implies independence. If distributions are not normal, zero correlation does not necessarily imply independence. Therefore, IID noise is white noise, but white noise may not be IID noise. In time series analysis, we assume the normality of residual distribution. Therefore, we consider that white noise satisfies the IID noise hypothesis in testing the randomness of residuals (see section 4C). In addition, white noise is enough to satisfy the no-autocorrelated-residuals assumption.[3] When we have figured out all the time-dependent behaviors in a time series, we will have white noise in most cases, and we are ready to forecast future values. In this respect, modeling a time series is the same as *whitening* the time series.

A time series with only a deterministic trend will be stationary after we eliminate the deterministic trend. This type of data is said to be *trend stationary*. A time series with a unit-root stochastic process will not be stationary even after we eliminate the deterministic trend.[4] This type of data is said to be *difference stationary* because we can eliminate the unit-root process in residuals by differencing them with an appropriate order.

We may be concerned about a trend, a long-term behavior of a time series, but not about its seasonality or the behavior of residuals. In this case, to make the trend stand out more clearly, we can reduce the short-term fluctuations in the time series without estimating them. This process is called *smoothing*. There are various techniques for this, such as exponential smoothing and moving average smoothing (see chapter 5).

When we test the causality from one time series to another time series, we need to whiten the two time series first, that is, remove any systematic patterns from the residuals of the two time series. Any observed relationship between two unwhitened time series may be biased. In the worst case, an observed relationship may simply be an artifact of the systematic patterns of the two series. That is, the two time series may tend to move in the same or opposite directions, even when neither of them is a cause of the other. This process of making the disturbances of time series variables consist of white noise before we analyze their relationship is called *prewhitening* (see chapter 6, section 2).

These are important concepts that are discussed in this book. When these and other concepts appear for the first time before they are fully explained in related sections or chapters, I briefly define them and tell readers where the concepts are discussed in detail.

## (2) PREPARING FOR ANALYSIS

### A. Data

In the social sciences, we try to provide our own generalizable explanations. As we need to be concerned with particulars to generalize our findings in cross-sectional analysis (King, Koehane, and Verba 1994), we need to determine carefully the length of period we will cover in time series analysis. When time series analysis covers too short a period, we may not generalize the findings. The observed trend, for example, may significantly change when we extend the period. In figure 2 (chapter 1, page 3), for example, if we consider only the 90th–96th Congresses, average leadership support scores on the party agenda will show an increasing trend both for the Republican party and for the Democratic party. In addition, if the number of time points is too small, time series analysis cannot be appropriately employed. In general, time series analysis requires at least 30 time points. If we analyze monthly crime rates, for example, we had better cover at least three years.

Extending the period of analysis, however, may not always be a good choice. If we extend the length of a time series too far, the chance that the time series consists of two or more parts that behave differently from each other increases. In addition, the chance that the nature of relationship between the time series and exogenous factors changes increases. If these are the cases, our estimate of the behavior of the time series and the forecast of its future values are likely to become less accurate. In analyzing the presidential vetoes in the United States from 1890 to 1994, for example, Brandt and Williams (2001) found a noticeable dynamic pattern in the data. Dividing the period into the pre- and post-FDR periods, however, they found that the noticeable dynamic pattern actually existed only during the post-FDR period (after 1945). Also, according to autocorrelation functions, some serial correlation was present in the presidential vetoes only during the post-FDR period.

In chapters 2–6, I analyze the monthly rates of violent crimes (murder, forcible rape, robbery, and aggravated assault) during the ten-year period from January 1983 to December 1992 as an example at the end of each step of the time series modeling procedure. The monthly violent crime rates in 1993 were not included in the modeling procedure. They were compared with the values forecasted by the estimated model, instead.

The period of analysis does not matter because I analyzed monthly violent crime rates as an example. I selected this period because violent crime was a continuing social disaster in the United States in the 1980s and 1990s. Approximately one-third (29%) of Americans thought that crime was a real problem in their neighborhood (Maguire, Pastore, and Flanagan 1993). Various forms of tougher crime-control policies, like the Three Strikes and You're Out law in California, were enacted in many states. Social scientists conducted various types of research on crime rate trends or on the impact of crime-control policies on crime reduction (see e.g. Ekland-Olson and Kelly 1992; Greenwood 1994; Stolzenberg and D'Alessio 1997; Zimring, Karnin, and Hawkins 1999).

According to violent crime rates per 100,000 inhabitants, extending the period back to 1970 is not likely to produce a significantly different trend estimate (figure 6). The model fit of the estimated trend line is likely to decrease slightly. I do not extend the period after 1994 because violent crime rates have continued to decrease since 1994. If we extend the period after 1994, the upward trend until 1994 and the downward trend after 1994 will cancel, making the least squares estimate of a linear trend inappropriate. In addition, the model fit of the estimated trend is likely to decrease.

I selected the period from 1985 to 2004, ten years before and ten years after the implementation of the Violent Crime Control and Law Enforcement Act (Public Law 103–322) in 1994, for the interrupted time series analysis of the getting-tougher-on-crime policy (see chapter 7).

The monthly violent crime rates from 1983 to 2004 were compiled with the data from *Uniform Crime Reports: Crime in the United States*, which is issued annually by the Federal Bureau of Investigation (see appendix 2). When the FBI reports did not directly provide monthly crime rates but monthly percentage distribution of crimes within each year, the monthly crime rates were calculated by multiplying the monthly percentage distribu-

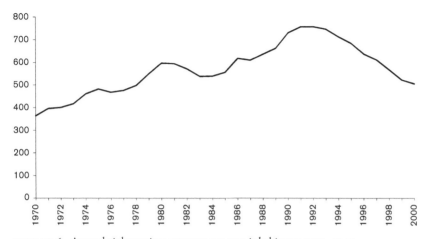

FIGURE 6. Annual violent crime rates per 100,000 inhabitants, 1970–2000.
SOURCE: U.S. Census Bureau, *Statistical Abstract of the United States, 1976, 1987, 1994, 2003* (Washington, DC: U.S. Government Printing Office)

tion and the corresponding yearly violent crime rate per 100,000 inhabitants. When the FBI reports did not include the violent crime rate total, the monthly violent crime rate total was the sum of the monthly crime rates for murder, forcible rape, robbery, and aggravated assault.

## B. Preliminary Visual Analysis

In univariate time series analysis where Time is the only independent variable, the focus is on the time-dependent behavior of a time series. The best way to gain a general idea of the time-dependent behavior of a time series is to visually inspect the time series. Thus, the first step of univariate time series analysis is to construct scatter plots with a time series on the vertical axis and Time on the horizontal axis and to visually examine the relationship between the time series and Time. Stand-alone time series analysis programs, as well as multipurpose statistical programs, provide an easy one-step procedure for the plotting of a time series.[5] Several points that we need to check for are

(a) a trend
(b) a sudden change in behavior
(c) periodic components
(d) smoothness

(e) the non-constant variability of noise

(f) outliers.

*Trend* is a tendency of a certain elemental characteristic to increase or decrease over time. In most cases, it is easy to determine through visual inspection whether a trend is present and, if it is present, whether it is linear or curvilinear, whether it increases or decreases as time passes, and whether it changes abruptly.

A time series may not contain any discernible trend (see e.g. figure 3, chapter 1, page 6). In OLS regression analysis, a slope which is not significantly different from zero is taken as evidence of no relationship between the dependent variable and an independent variable. In time series analysis, a trend that is not significantly different from zero does not necessarily provide evidence of no significant relationship between a time series and Time. A time series with no discernible trend can still show a time-dependent behavior due to its residuals with a systematic pattern.

*Periodic components* are often present when a time series consists of seasonal, for example monthly or quarterly, observations. When they are present, it will not be difficult to identify a cyclical periodic pattern in the scatter plots of raw observations. We need to remove periodic components from a time series to accurately estimate other time-dependent behaviors, such as trends and systematic patterns of residuals (see section 3).

*Smoothness* of the scatter plots of a time series, essentially long stretches of plots on the same side above or below the trend line, indicates the presence of a systematic pattern of residuals. Systematic patterns of residuals consist of an AR process and/or an MA process. These processes can be estimated after both trend and seasonal components are estimated and then removed from a time series. Trend and/or seasonal components can also be eliminated from a time series by differencing the time series (see section 4).

If *non-constant variance* of a time series, known as *heteroscedasticity* in OLS regression analysis, is obvious in the scatter plots, the time series needs to be transformed before its time-dependent behaviors are estimated (see section 2C). Non-constant variance may not consistently increase or decrease but erupt in some periods. In this case, we need to respecify our model or model this *conditional heteroscedasticity* (see section 5B).

*Outliers* are observations that have extremely large residuals. They can be detected by examining the scatter plots of raw observations against Time, the residual plots against Time, and the residual plots against the original time series. Outliers may not distort the trend estimation. That is, we may have similar trend estimates with and without outliers. Then we may not need to care much about outliers. If they distort trend estimates, that is if they clearly increase or decrease trend estimates or if they change the direction of trend estimates, we should carefully determine how to deal with outliers.

Outliers may turn out to be normal with additional observations, but it is difficult to collect additional past observations of a time series. Outliers may be cases whose values were incorrectly recorded. Then, we simply need to correct them. We may clarify a particular reason for which outliers exist at particular time points, such as the sudden increase in the death rate in a country when the country waged war against another country. Then, we can include a dummy variable scheme to control out the particular event, creating an independent variable that assigns 1 to the particular time points and 0 to the others.

If outliers apparently distort trend estimates but we cannot clarify why they exist, we need to conduct two separate analyses, with and without outliers, and report both results. Excluding outliers is not as easy in time series analysis as in OLS regression analysis of cross-sectional data. A time series has only one observation at each time point, and therefore excluding outliers will cause a missing-data problem. We may replace the excluded values with *imputed* or *interpolated* values, which are likely to be mixed better with other observations than the outliers (Honaker and King 2010; King et al. 2001; Little and Rubin 2002). Briefly, in the imputation approach, we retrieve relevant information from the available data and impute multiple values for each missing cell to construct multiple datasets with no missing values. We analyze these completed datasets and then combine the results. In the interpolation approach, we replace the missing values with interpolated values. There are several interpolation algorithms, such as linear, polynomial, and spline.

Figure 7 presents a time series plot of the monthly violent crime rates from 1983 to 1992. In this figure, we can roughly figure out that the time series consists of a linear trend and seasonal components of lag 12, and possibly systematically patterned residuals. In most cases, a linear model will fit a time series well. We may try other models, such as quadratic and exponential models, to determine which best fits the time series. When plotting a time series, most statistical programs also report an estimated trend. The monthly violent crime rates from January 1983 to December 1992 appear to contain an upward linear trend.

Figure 7 shows a seasonal pattern, with a peak in July or August and a trough in February (see also figure 8). This seasonality is theoretically plausible in that crimes are likely to be committed more frequently during warm

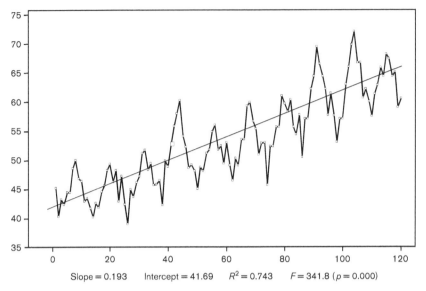

Slope = 0.193     Intercept = 41.69     $R^2$ = 0.743     $F$ = 341.8 ($p$ = 0.000)

FIGURE 7. Monthly violent crime rates in the US, January 1983 to December 1992.

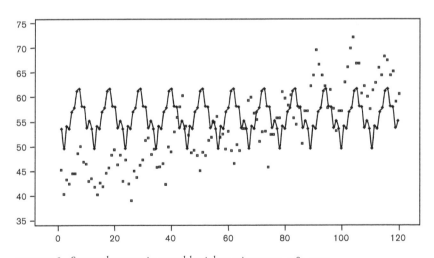

FIGURE 8. Seasonal pattern in monthly violent crime rates, 1983–1992.

periods, especially around summer, and less frequently during cold periods. Figure 7 also shows smoothness. This smoothness indicates that AR and/or MA processes may be present in the residuals. Because trend and seasonal components are present, however, we first need to remove them to determine with more confidence whether the residuals contain a systematic pattern.

Visually inspecting sample autocorrelation function (ACF) correlograms can also help us gain a general idea of the time-dependent behavior of a time series (see section 5). When a time series contains a trend, its ACFs slowly decay because current observations are significantly related to previous observations. An example of slowly decaying ACFs is shown in figure 9. If periodic components are present, the slowly decaying ACFs will additionally show a cyclical pattern (see figure 10).

The visual inspection of sample autocorrelations of the monthly violent crime rates confirms that this time series consists of a linear trend and seasonal components of lag 12. In figure 11, ACFs slowly decay over time, indicating the presence of a linear trend in the time series. Also, ACFs reach the highest point at every 12th lag, indicating that the time series contains seasonal components of lag 12. When producing correlograms, statistical programs report the numerical values of ACFs. Thus, we can also refer to these values to determine the pattern of periodic components. Since the time series in figure 11 contains a trend and seasonal components, we cannot tell whether or not its residuals contain a systematic pattern.

## C. Data Transformation

The variance of a time series may not be close to constant for all the values of the independent variable, Time. It may increase (or decrease) with the level of the process, instead. In OLS regression analysis, an increasing or decreasing variance of the dependent variable with an increasing level of an independent variable is termed *heteroscedasticity*. With the presence of heteroscedasticity, the estimated OLS coefficient is still unbiased but the size of its standard error is influenced. When the variance of observations is not constant across time, the estimated trend that is the OLS estimate will be unbiased, but its standard error will be influenced. As a result, the estimated trend may not be the best among a class of unbiased estimates.

If we detect heteroscedasticity in the scatter plots of time series data, we can employ statistical tests to check the seriousness of non-constant variance. For instance, the Goldfield-Quandt test divides the dependent variable into

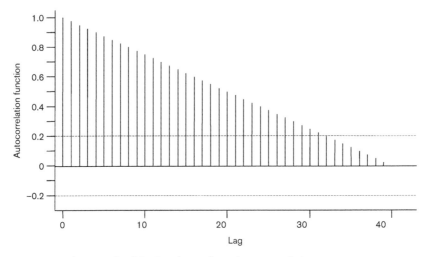

FIGURE 9. An example of the slow decay of sample autocorrelations.

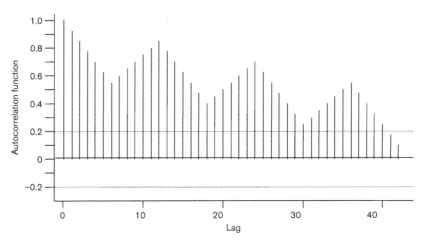

FIGURE 10. An example of the slow decay of sample autocorrelations with periodic components.

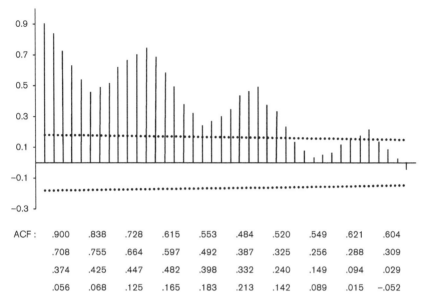

FIGURE 11. Sample autocorrelation function (ACF) of monthly violent crime rates, lags 1–40.

three parts according to the temporal order and fits the OLS regression line for the first and third parts (Goldfield and Quandt 1965). Then it calculates the residual variance for the first part and the third part, and checks whether these two residual variances are significantly different from each other by using the $F$ statistic. In this case, the degree of freedom of the $F$ statistic in the numerator and the denominator will be equal to

$$\frac{n - n_2}{2} - 2$$

where $n$ is the total number of observations and $n_2$ is the number of observations of the second part.

In time series data, however, there is only one observation for each time point, and the number of cases is usually small. Thus, in revealing the dependence of variability on the level of a time series, visual inspection of residuals is generally more appropriate than statistical tests.

When the variance is not constant across time, we can employ the weighted least squares approach or the Box-Cox transformation approach. The weighted least squares estimation would increase the precision of estimates, as in OLS regression analysis of cross-sectional data.[6] The weighted least squares approach, however, may be less appropriate in time series analysis than in OLS regression analysis of cross-sectional data. Each value of Time or an independent variable has only one corresponding value of the dependent time series, while we have multiple observations of the dependent variable per each value of an independent variable in cross-sectional analysis.

In time series analysis, the Box-Cox transformation is frequently employed to stabilize residual variance (Box and Cox 1964). It takes natural logarithms of the original series, $\ln(X_t)$, when the variance (standard deviation) increases almost linearly. In other cases, the transformation equation is

$$X_{t(\lambda)} = \frac{X_t^{\lambda} - 1}{\lambda}$$

where $X_t \geq 0$ and $\lambda \neq 0$.

If statistical programs do not provide the Box-Cox transformation procedure, we can utilize the logarithmic transformation method to stabilize a linearly increasing variance. In other cases, we can directly transform a time series with any value of $\lambda$ by using a statistical program's compute procedure. Often, 0.5 is an adequate value for $\lambda$. The best value can be determined by trial and error. We can determine an adequate value for $\lambda$ by visually inspecting the transformed series.

Non-constant variance of a time series may not be worth correcting every time it is detected. First, the impact of non-constant variance on the size of the standard error of a trend estimate will depend on the sample size, the range of the values of the independent variable, and the relationship between the non-constant variance and the independent variable (Berry and Feldman 1985; Fox 1991). As the sample size becomes larger or as the interval between the smallest value and the largest value of the independent variable increases, the problem of non-constant variance will diminish. The sample size of a time series dataset is usually small. The range of the independent variable, however, is usually wide. In the monthly violent crime rates from 1983 to 1992, for example, the minimum of the independent variable, Time, is 1, and the maximum is 120, although its number of cases is 120. Thus, its range is wider than most independent variables in cross-sectional analysis that are measured by using three-to-seven-point scales.

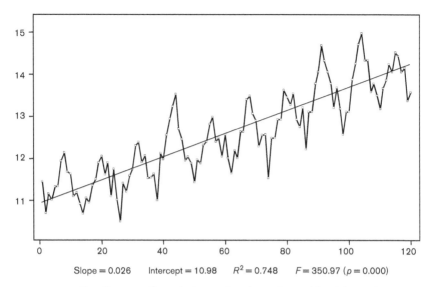

Slope = 0.026    Intercept = 10.98    $R^2$ = 0.748    $F$ = 350.97 ($p$ = 0.000)

FIGURE 12. Box-Cox transformed time series plots of monthly violent crime rates, 1983–1992 ($\lambda$ = 0.5).

Second, transformation of an original time series may not always help us correct non-constant variance. If variance does not consistently increase or decrease with the level of the process, transformation of the time series will not have a significant effect on the non-constant variance. In figure 7, for example, variance does not consistently increase. It actually repeats, in an increase-and-decrease pattern. Box-Cox transformation of figure 7 reduces the range of crime rates, but it does not bring about any significant change of noise fluctuations (see figure 12). The estimated trend was statistically significant both before and after transformation. The direction of the estimated coefficient was identical and the model fit remained mostly unchanged before and after transformation. The only noticeable change was that the estimated trend was damped after transformation because the range of the dependent variable was reduced.

In addition, once we transform a time series, we should keep in mind that we model the transformed series, not the original series. Our trend estimates will have least square properties with respect to the transformed series and not to the original series. Therefore, our interpretation of the model will not contain an intuitive meaning of the relationship between Time and the dependent variable. In addition, the predicted values will not pertain to the original series. They will be the predicted values of the transformed series.

## (3) SEASONAL COMPONENTS AND TREND

The time series modeling procedure consists of three separate sub-procedures, which are respectively designed to estimate and then eliminate seasonality, trends, and systematic patterns of residuals in order. The first two procedures are explained in this section. The third procedure is explained in the next section.

### A. Deseasonalization

For a time series with no apparent periodic components, we can conduct bivariate regression analysis, in which Time is the independent variable, to estimate a trend and obtain residuals. However, periodic regularity may be present in a time series that contains seasonal, for example monthly or quarterly, observations. This periodic regularity, along with a trend and a systematic pattern of residuals, characterizes the behavior of the time series. In this case, seasonal components will mask a trend and a systematic pattern of residuals. Therefore, estimating and eliminating seasonal components is the first step we need to take to accurately estimate the time-dependent behavior of the time series.

The procedure of estimating and eliminating seasonal components is termed the *seasonal adjustment* or *deseasonalization* of a time series. In most cases, seasonal components are easily detected by visually inspecting a time series and its sample autocorrelations (see section 2). Once detected, it can be easily estimated and then eliminated from a time series. To perform this task, we need to determine the length of the observed cyclical period. This is a simple task in most cases: the length will be 12 lags in a monthly time series and 4 in a quarterly time series. The length of a cyclical period will be detected by visually inspecting an original series or its sample ACFs. In figures 7 and 11, for example, the monthly violent crime rates from 1983 to 1992 and the sample ACFs reach their highest points at every 12th lag, indicating that each cyclical period lasts for 12 lags.

Seasonality is estimated through a two-stage procedure. First, if an obvious trend is present in a time series, it is estimated and removed. Because this estimation is done without removing seasonality, the estimated trend is not final and may be inaccurate. This is known as a *pre-estimated trend*. The purpose of estimating and removing it is to estimate seasonality accurately. Second, with the detrended series, we estimate the seasonal components. After we estimate and remove the seasonal components from the original series, we are ready to re-estimate the real trend, as explained below. Most statistical programs provide an easy one-step procedure of deseasonalization, and we are not required to conduct the first procedure.[7]

Seasonal components can be estimated in two different ways: the additive method and the multiplicative method. These two methods lead to mostly identical results unless the range of seasonal components is obviously proportional to the level of a time series. If the range of seasonal components tends to increase or decrease obviously over time, it is better to employ the multiplicative method.

Seasonal components that are estimated by the *additive method* describe how much each seasonal component pulls up or down the overall level of a time series, on average. A time series without seasonal variation has a seasonal component of zero. When seasonal components are estimated by the additive method, we deseasonalize a time series by subtracting the seasonal components from the original series. Thus, a seasonally adjusted series will decrease when a seasonal component pulls up the original time series and increase when a seasonal component pulls down the original series. To restore the original time series from a seasonally adjusted series, we add seasonal components to the seasonally adjusted series.

Seasonal components that are estimated by the *multiplicative method* are factors by which a time series is divided to yield a seasonally adjusted series. These factors are the ratios by which the overall level of a time series increases or decreases, on average. A time series without seasonal variation has a seasonal component of one. When we employ the multiplicative method, we deseasonalize the original time series by dividing it by the estimated seasonal components. Thus, when a seasonal component increases (decreases) the original time series by a certain ratio, the seasonally adjusted series will decrease (increase) by that ratio. To restore the original time series from a seasonally adjusted series, we multiply the seasonally adjusted series by the seasonal components.

As observed in figures 7 and 11, the monthly violent crime rates from 1983 to 1992 contain seasonal components of lag 12. As its trend is compounded with seasonality, we need to estimate and remove the seasonal components for an accurate estimation of a trend. The estimated seasonal components of the monthly violent crime rates from 1983 to 1992 are presented in table 1.

According to the seasonal components estimated by the additive method, the seasonal component for February tends to pull down the overall crime rate by 6.93 points on average during the ten-year period included in the time series modeling. The seasonal component for August tends to pull up the overall crime rate by 5.86 points on average. According to the seasonal components that are estimated by the multiplicative method, the seasonal

TABLE I. Estimated Seasonal Components of the Monthly
Violent Crime Rates, 1983–1992

| Month | Seasonal components | |
|---|---|---|
| | Additive method | Multiplicative method |
| January | –2.72 | .95 |
| February | –6.93 | .87 |
| March | –2.16 | .96 |
| April | –2.69 | .95 |
| May | .90 | 1.02 |
| June | 1.79 | 1.03 |
| July | 5.39 | 1.10 |
| August | 5.86 | 1.11 |
| September | 2.15 | 1.04 |
| October | 1.96 | 1.04 |
| November | –2.51 | .95 |
| December | –1.05 | .98 |

component for February tends to decrease the overall level of the series by a ratio of 0.87 on average during the ten-year period included in the time series modeling. The seasonal component for August tends to increase the overall level by a ratio of 1.11 on average.

As explained, we subtract the seasonal components from the original time series to deseasonalize the time series, when we estimated seasonal components by the additive method. When we estimated them by the multiplicative method, we divide the original series by the seasonal components. We can let statistical programs report the seasonal components and save a seasonally adjusted series. Table 2 shows the deseasonalization procedure for the monthly violent crime rates in 1990.

Because seasonal components were removed from the original series, the deseasonalized series was damped (figure 13). As shown in figure 7, the range of seasonal components does not clearly increase or decrease over time. Thus, the time series seasonally adjusted by the additive method and the time series adjusted by the multiplicative method are nearly identical (figure 14).

## B. Trend Estimation

Estimating trends in univariate time series analysis is analogous to estimating slopes in bivariate OLS regression analysis. We fit the least squares line to a

TABLE 2. Deseasonalization of the 1990 Monthly Violent Crime Rates

| | | Additive method | | Multiplicative method | |
| | Original series | Seasonal component | Adjusted series | Seasonal component | Adjusted series |
|---|---|---|---|---|---|
| January | 57.81 | –2.72 | 60.53 | .95 | 60.85 |
| February | 50.49 | –6.93 | 57.41 | .87 | 58.03 |
| March | 57.08 | –2.16 | 59.24 | .96 | 59.46 |
| April | 57.08 | –2.69 | 59.77 | .95 | 60.08 |
| May | 62.20 | .90 | 61.30 | 1.02 | 60.98 |
| June | 64.40 | 1.79 | 62.61 | 1.03 | 62.52 |
| July | 69.52 | 5.39 | 64.13 | 1.10 | 63.20 |
| August | 66.59 | 5.86 | 60.73 | 1.11 | 59.99 |
| September | 64.40 | 2.15 | 62.25 | 1.04 | 61.92 |
| October | 62.20 | 1.96 | 60.24 | 1.04 | 59.81 |
| November | 57.81 | –2.51 | 60.32 | .95 | 60.85 |
| December | 61.47 | –1.05 | 62.52 | .98 | 62.72 |

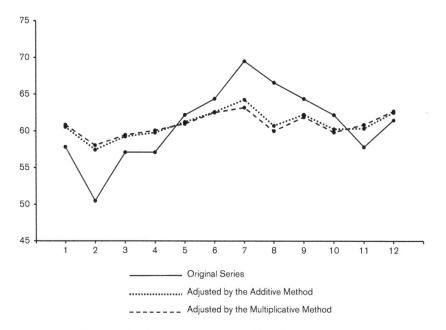

———— Original Series

·············· Adjusted by the Additive Method

– – – – – – Adjusted by the Multiplicative Method

FIGURE 13. Deseasonalized time series: 1990 monthly violent crime rates.

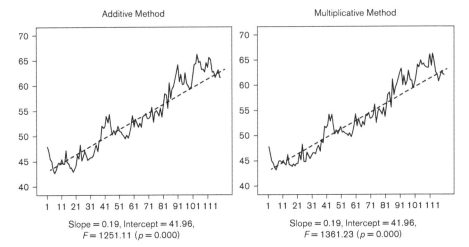

Additive Method  Multiplicative Method

Slope = 0.19, Intercept = 41.96,
F = 1251.11 (p = 0.000)

Slope = 0.19, Intercept = 41.96,
F = 1361.23 (p = 0.000)

FIGURE 14. Time series plots of monthly violent crime rates, 1983–1992, deseasonalized by the additive or multiplicative method.

time series, just as we do in bivariate OLS regression analysis. When estimating a trend, it is helpful to visually inspect a time series. We can determine whether there is a discernible trend, whether it is increasing or decreasing, and whether it is linear or nonlinear.

When a trend is nonlinear, the order of a polynomial trend can be determined according to the number of bends in the trend line: the former will be greater than the latter by one (Berry and Feldman 1985). A linear trend is a polynomial trend of order one. In the social sciences, a time series is unlikely to have a polynomial trend of an order higher than three. If it is difficult to determine the order of a polynomial trend, we can try one by one among a linear trend $(\beta_0 + \beta_1 \text{Time})$, a quadratic trend $(\beta_0 + \beta_1 \text{Time} + \beta_2 \text{Time}^2)$, and a cubic trend $(\beta_0 + \beta_1 \text{Time} + \beta_2 \text{Time}^2 + \beta_3 \text{Time}^3)$. If a trend is curvilinear but continues to increase or decrease without a bend, we can try a power trend $(\beta_0 \times \text{Time}^{\beta_1})$, a growth trend $(e^{\beta_0 + \beta_1 \text{Time}})$, an exponential trend $(\beta_0 e^{\beta_1 \text{Time}})$, and a hyperbolic trend $(\beta_0 + \beta_1 \frac{1}{\text{Time}})$. We can check which of them best fits the time series by inspecting the residual plots and the model fits (see section 5, chapters 3 and 4). When a time series contains a polynomial trend with one or more bends, we may consider dividing the polynomial trend into two or more parts of positive and negative linear trends.

After removing the seasonality from the monthly violent crime rates from 1983 to 1992, we can clearly identify an upward linear trend in the plot of

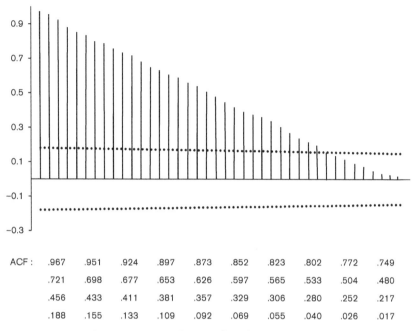

| ACF : | .967 | .951 | .924 | .897 | .873 | .852 | .823 | .802 | .772 | .749 |
|-------|------|------|------|------|------|------|------|------|------|------|
|       | .721 | .698 | .677 | .653 | .626 | .597 | .565 | .533 | .504 | .480 |
|       | .456 | .433 | .411 | .381 | .357 | .329 | .306 | .280 | .252 | .217 |
|       | .188 | .155 | .133 | .109 | .092 | .069 | .055 | .040 | .026 | .017 |

FIGURE 15. Sample autocorrelation function (ACF) of deseasonalized monthly violent crime rates, lags 1–40.

the deseasonalized series (figure 14) and its ACF correlogram (figure 15). By fitting the least squares line to the deseasonalized series, we obtain the following linear equation:

$$M_t = 41.96 + 0.19 \times \text{Time}$$

where Time = 1, 2, ..., 120.

The standard errors of the trend estimate and the constant are 0.005 ($t = 35.37$) and 0.37 ($t = 112.94$), respectively.

## (4) SYSTEMATIC PATTERNS OF RESIDUALS

We have estimated the trend and the seasonal components in the monthly violent crime rates. The line that has been fitted by these two estimates is marked in figure 16. As shown in the figure, the original data are well represented by the estimated trend and seasonality. There are, however, many observations that are

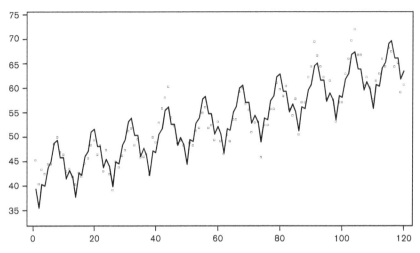

FIGURE 16. Fitted line by the trend and seasonality estimates.

not exactly on the fitted line. As in OLS regression analysis, the distance between each observation and the fitted line is a *residual*.

In a time series, residuals often contain systematic temporal patterns that are left unexplained by the estimated trend and seasonality. In time series analysis, the two most frequently identified systematic patterns of residuals are an *AR process* and an *MA process*. These are what we should remove from residuals to whiten them. When an AR process exists, residuals at different time points are correlated with each other. Therefore, residual plots will show smoothness: long stretches of plots on the same side above or below the trend line. An AR process is frequently observed in a time series that is related to human behavior, economic conditions, or social phenomena. Unemployment rates, for instance, tend to continuously improve and then continuously worsen, rather than randomly fluctuating, even after we deseasonalized and detrended them. Such a tendency is not influenced by exogenous factors. Rather, the values of residuals just tend to systematically change. Therefore, when an AR process is present, we just model the process rather than figuring out the cause of the serial correlation.

An MA process depicts some kind of systematically repeated pattern, which can be estimated by the average frequency of the repeated movement. An MA process is more common in natural phenomena than in social phenomena. Volcanic activities, for instance, may tend to be repeated every 100 years on average. A strong earthquake may tend to strike back every 70 years on average. In the social sciences, an MA process is usually estimated to account for a systematic pattern that remains after an AR component was removed from the residuals (see e.g. Beck 1991; Green, Gerber, and De Boef 1999; Green, Palmquist, and Schickler 2001).

## A. How to Obtain Residuals

We can obtain residuals either by the estimate-and-then-eliminate procedure or by the differencing procedure. If a time series contains a trend and seasonality, we first estimate seasonality and save the deseasonalized series. Then, when fitting a trend line for the deseasonalized series, we save the residuals, that is the deseasonalized and detrended series (hereafter, $\{Y_t\}$). Most programs automatically save residuals, and thereby we have residuals ready to be estimated after we estimated seasonality and a trend. Figure 17 presents the plots of the deseasonalized and detrended monthly violent crime rates that were produced through the estimate-and-then-eliminate procedure.

In time series analysis, another convenient way to produce residuals is *lag*-k *differencing* ($\nabla_k$). It directly removes a trend and/or seasonality from a time series. In lag-$k$ differencing, the lag of differencing, $k$, indicates that the $k$th previous value is used to calculate the difference. That is, the lag-$k$ differencing of a time series is equivalent to converting every $(i + k)$th value of the time series into its difference from the $i$th value.

The lag of differencing is determined according to the characteristic of the non-stationary process. For example, if a time series contains only a linear trend ($X_t = \beta_0 + \beta_1$ Time $+ Y_t$), the lag-1 differencing will remove the trend of the time series, $\beta_1$, and leave only residuals:

(1)  $\nabla_1 X_t = X_t - X_{t-1}$

$= (\beta_0 + \beta_1 \text{ Time} + Y_t) - \{\beta_0 + \beta_1 (\text{Time} - 1) + Y_{t-1}\}$

$= \beta_0 + \beta_1 \text{ Time} - \beta_0 - \beta_1 \text{ Time} + \beta_1 + (Y_t - Y_{t-1})$

$= \beta_1 + (Y_t - Y_{t-1})$

In equation (1), the time-dependent component, $\beta_1$ Time, is removed. If the detrended series, $\{Y_t\}$, contains a constant mean equal to zero, we will have a transformed time series with a mean equal to $\beta_1$.

When a time series contains monthly seasonal components, the lag of differencing will be 12. The lag-12 differencing subtracts the seasonal component of each month of the previous year from the seasonal component of

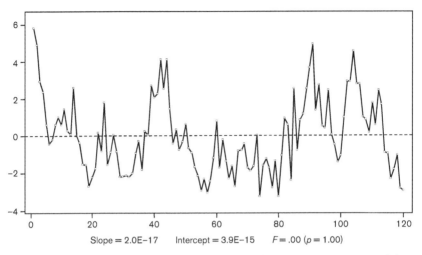

Slope = 2.0E–17    Intercept = 3.9E–15    F = .00 (p = 1.00)

FIGURE 17. Deseasonalized and detrended monthly violent crime rates, 1983–1992, $\{y_t\}$.

the same month of the current year, that is the 1st value from the 13th value, the 2nd value from the 14th value, and so on:

$$(2) \qquad \nabla_{12} S_t = S_t - S_{t-12} = 0$$

To remove quarterly seasonal components, the lag of differencing is 4: we subtract the 1st value from the 5th value, the 2nd value from the 6th value, and so on.

The lag of differencing should not be confused with the *order of differencing*, which is usually denoted by $d$. The order of differencing is the number of times we difference a time series to obtain stationary residuals. In this case, this time series is said to be *integrated of order d*, denoted by $I(d)$. For a time series that contains only a linear trend, doing the lag-1 differencing once will remove the trend. Therefore, the order of differencing is also 1.

If we have a time series with a quadratic trend, the order of differencing will be 2. We do the lag-1 differencing twice. We first remove a quadratic trend by applying the lag-1 differencing to the time series (equation (3)). Then, by applying the lag-1 differencing to the lag-1 differenced series, we remove a linear trend (equation (4)). We will have a transformed time series with a mean equal to 2 $\beta_2$, where $\beta_2$ is the coefficient of the quadratic term, $Time^2$, in the original series:

(3) $\nabla_1 X_t = \nabla_1(\beta_0 + \beta_1 \text{Time} + \beta_2 \text{Time}^2) + \nabla_1 Y_t$

$\quad = (\beta_0 + \beta_1 \text{Time} + \beta_2 \text{Time}^2) - \{\beta_0 + \beta_1 (\text{Time} - 1) +$
$\quad \beta_2 (\text{Time} - 1)^2\} + (Y_t - Y_{t-1})$

$\quad = (\beta_1 + 2\beta_2 \text{Time} - \beta_2) + (Y_t - Y_{t-1})$

(4) $\nabla_1\{(\beta_1 + 2\beta_2 \text{Time} - \beta_2) + (Y_t - Y_{t-1})\}$

$\quad = \{\beta_1 + 2\beta_2 \text{Time} - \beta_2 + (Y_t - Y_{t-1})\} - \{\beta_1 + 2\beta_2 (\text{Time} - 1) - \beta_2 +$
$\quad (Y_{t-1} - Y_{t-2})\}$

$\quad = \beta_1 + 2\beta_2 \text{Time} - \{\beta_1 + 2\beta_2 (\text{Time} - 1)\} + \{(Y_t - Y_{t-1}) - (Y_{t-1} - Y_{t-2})\}$

$\quad = 2\beta_2 + (Y_t - 2Y_{t-1} + Y_{t-2})$

In the social sciences, a trend with two or more bends is unusual, and therefore we are not likely to try an order of differencing greater than 2 just to eliminate trends. In addition, when we observe a trend with two or more bends in a time series, we may need to divide the time series into two or more subsets and apply the lag-1 differencing to each subset separately.

The lag-$k$ differencing reduces the number of cases by $k$ because the difference cannot be calculated for the first $k$ cases. For example, the first case of the lag-1 differenced series, in which a linear trend was eliminated, is the difference between the first and second cases in the original series. The first case of the original series has the system-missing value for the lag-1 differenced series. The first 12 cases of the original series have the system-missing values for the lag-12 seasonal differenced series.[8]

Lag-$k$ seasonal differencing will also remove a linear trend from a time series. For a monthly linear time series, for example, we will have a transformed time series with a mean equal to 12 $\beta_1$ and without a linearly time-dependent component $\beta_1$ (Time):

(5) $\nabla_{12} X_t = \nabla_{12}(\beta_0 + \beta_1 \text{Time}) + \nabla_{12} S_t + \nabla_{12} Y_t$

$\quad = [\beta_0 + \beta_1 \text{Time} - \{\beta_0 + \beta_1(\text{Time} - 12)\}] + (S_t - S_{t-12}) + (Y_t - Y_{t-12})$

$\quad = 12 \beta_1 + (Y_t - Y_{t-12})$

Figure 18 shows that the lag-12 seasonal differencing (or the order-1 seasonal differencing) removed the trend of the monthly violent crime rates from 1983 to 1992: the slope is not significantly different from zero, and the regression equation is insignificant.

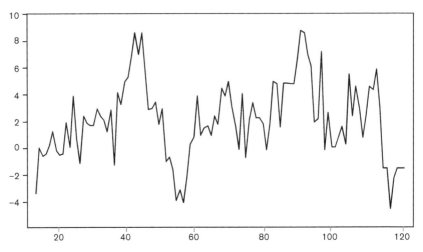

FIGURE 18. Lag-12 differenced monthly violent crime rates, 1983–1992.

Although we can directly produce residuals by differencing a time series, the estimate-and-then-eliminate method is preferred to the differencing method in social science studies for several reasons. First, in most social science studies, we need to estimate a trend and seasonality. When we difference a time series, we may estimate its trend, but not always. When a time series contains only a linear trend, we can estimate the trend by calculating the mean of the lag-1 differenced series, if $\{Y_t\}$ has a mean of zero (see equation (1)). For a quadratic series, we can calculate the coefficient of a quadratic term (see equation (4)) but not the coefficient of a linear term (see equation (3)).

Second, when we use the estimate-and-then-eliminate method, we can keep all the cases in the original time series. When we apply lag-$k$ differencing to a time series, the number of available cases is reduced by 1 when we remove a linear trend, by 4 when we remove quarterly seasonal components, by 12 when we remove monthly seasonal components, and so on. For this reason, the estimate-and-then-eliminate method is preferred to the differencing method especially when the number of cases is small, which is often the case in social science time series analysis.

Third, residuals may behave differently when they are produced by the estimate-and-then-eliminate method versus when they are produced by the differencing method. For example, by estimating and then removing a trend from a non-seasonal time series, $\{\beta_0 + \beta_1 \text{ Time} + Y_t\}$, we will have $\{Y_t\}$ as residuals. By applying the lag-1 differencing to the same series, we will have $\{Y_t$

$- Y_{t-1}$} as residuals (see equation (1)). If $\{Y_{t-1}\}$ and $\{Y_t\}$ are significantly correlated with each other, $\{Y_t - Y_{t-1}\}$ and $\{Y_t\}$ will behave differently. For example, if $Y_t = 0.7Y_{t-1} + Z_t$, then $Y_t - Y_{t-1} = 0.7Y_{t-1} - Y_{t-1} + Z_t - Z_{t-1} = -0.3Y_{t-1} + Z_t - Z_{t-1}$. As in this case, differencing an AR(1) process produces a new ARMA(1, 1) process with a different AR coefficient and an MA coefficient of 1.[9] A related issue is *over-differencing*. If we difference a time series such as white noise that does not need further differencing, the randomness of residuals may decrease. For example, if $\{Y_t\}$ is white noise, differencing $\{Y_t\}$ may cause a new systematic pattern in the residuals characterized by excessive changes of sign from one residual to the next and also by an increased standard error.

An AR process of order greater than 2 is rare in the social sciences. Therefore, we are not likely to have differently behaving residuals when we employ the estimate-and-then-eliminate method and when the lag of difference is greater than 2, that is, when we seasonally difference a time series that contains quarterly or monthly observations. For example, the deseasonalized and detrended monthly violent crime rates, $\{Y_t\}$, contains an AR process of order 2, meaning that the two adjacent past observations, $Y_{t-1}$ and $Y_{t-2}$, are significantly correlated with the current observation, $Y_t$.[10] Since $\{Y_t\}$ and $\{Y_{t-12}\}$ are not correlated with each other, the lag-12 differencing, with which we remove the seasonality from the monthly violent crime rates, will not influence the systematic pattern of residuals. That is, the systematic pattern of $\{Y_t - Y_{t-12}\}$ will not be significantly different from that of $\{Y_t\}$.

In fact, in the case of the monthly violent crime rates from 1983 to 1992, the estimate-and-then-eliminate method and the lag-12 differencing method produced similarly behaving residuals (figures 17 and 18). $\{Y_t\}$ and $\{Y_t - Y_{t-12}\}$ actually contain an identical AR process of order 2. For $\{Y_t\}$, the coefficient of the AR process at lag 1, $\phi_1$, is 0.54, and the coefficient of the AR process at lag 2, $\phi_2$, is 0.29. For $\{Y_t - Y_{t-12}\}$, $\phi_1$ is 0.50 and $\phi_2$ is 0.29. The difference between figures 17 and 18 is that plots begin from the 13th mark of the horizontal axis in figure 18 because we applied lag-12 differencing to the series. In addition, figure 17 depicts $\{Y_t\}$ and figure 18 depicts $\{Y_t - Y_{t-12} + 12\,\beta_1\}$.

If we apply lag-1 differencing to $\{Y_t - Y_{t-12}\}$, we have

$$\nabla_1(Y_t - Y_{t-12}) = \{(Y_t - Y_{t-12}) - (Y_{t-1} - Y_{t-13})\} = \{(Y_t - Y_{t-1}) - (Y_{t-12} - Y_{t-13})\}.$$

Since significant correlation exists between $\{Y_t\}$ and $\{Y_{t-1}\}$ and between $\{Y_{t-12}\}$ and $\{Y_{t-13}\}$, residuals in figures 17 and 19 show clearly different behavioral patterns. $\{Y_t\}$ in figure 17 contains an AR(2) process, while $\{(Y_t - Y_{t-1}) - (Y_{t-12} - Y_{t-13})\}$

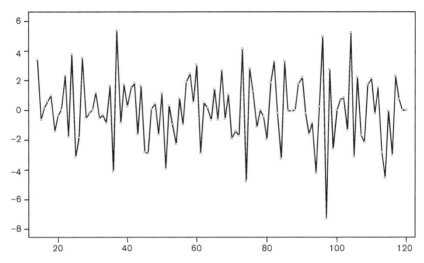

FIGURE 19. Lag-12 and then lag-1 differenced monthly violent crime rates, 1983–1992.

in figure 19 does not. In addition, plots begin from the 14th mark of the horizontal axis in figure 19 because we applied lag-12 differencing and then lag-1 differencing to the series.

## B. Visual Inspection

After removing a trend and seasonality from a time series, we need to check whether only white noise is left. To this end, we need to inspect residual plots. Smoothness in the residual plots, that is, long stretches of plots on the same side above or below the mean, indicates the presence of some type of systematic pattern in the residual behavior.

We need to inspect sample ACFs and partial autocorrelation functions (PACFs) to identify the type of systematic pattern in the residual behavior (see Section 5A). *Autocorrelation* at lag $k$ is correlation between a time series and its $k$th lag. *Partial autocorrelation* at lag $k$ is correlation between a time series and its $k$th lag after the correlations at intermediate lags have been partialled out. We calculate these correlations usually for the first 20–40 cases. This is why we call them sample autocorrelation and sample partial autocorrelation.

Once a trend and periodic components are removed from a time series, ACFs will no longer slowly decay. In most cases, the abrupt decay of ACFs, that is, the roughly geometric decay of the first few sample autocorrelations,

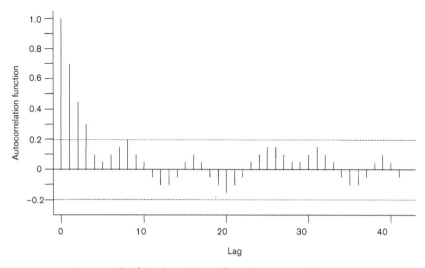

FIGURE 20. An example of the abrupt decay of sample autocorrelations.

suggests the presence of an AR process in a deseasonalized and/or detrended time series. The abrupt decay of PACFs is an indicator of an MA process. An example of abruptly decaying ACFs is shown in figure 20.

To determine more accurately whether AR and/or MA processes are present, we need to check how many bars cross the confidence bounds in ACF and PACF correlograms (see section 5A). When statistical programs produce sample ACF and PACF correlograms, they mark the 95% confidence bounds:

$$\pm 1.96 \, / \, \sqrt{n}$$

where $n$ is the number of observations.

When there is time dependency in a time series, and thus, when standard errors are different across ACFs and PACFs, the 95% confidence bounds are calculated by multiplying the standard errors of ACFs and PACFs by ±1.96, which is the $Z$ value at $\alpha = .05$. In this case, the confidence bounds will change as the lag increases.

In sample ACF and PACF correlograms, the ACF and PACF at lag zero lie outside the confidence bounds, reaching 1, because they are correlations between the same time series. For this reason, they are not usually included in the correlograms. If no systematic pattern exists in the residuals, that is, if the residuals are a realization of an IID or white sequence, we expect that about 95% of sample ACFs at lag 1 or higher will fall inside the 95% confidence bounds. If more than 5% of ACFs other than the first one at lag zero cross the confidence bounds, and if they show rapid decay, as in figure 20, we need to think of the presence of some type of AR process in the residuals. If more than 5% of PACFs cross the bounds and if they show rapid decay, we need to think of the presence of some type of MA process.[11] We have 120 observations in the crime rates data. Thus, we expect that less than six sample ACFs will fall outside the bounds ($\pm 1.96 / \sqrt{120} = \pm 0.179$), if the residuals are a realization of an IID or white sequence.

Figure 17 plots the residuals that were obtained by eliminating the estimated seasonality and trend from the original crime rates (figure 7). Although there is no discernible trend or seasonality, the smoothness of the graph clearly indicates that some form of dependency exists between residuals across time. We can confirm this point by inspecting the sample ACF and PACF correlograms of the residuals (figures 21 and 22). In figure 21, ACFs abruptly decay and ACFs at lags 1–4 and 24–32 lie outside the 95% confidence bounds. Since 12 out of 40 sample ACFs fall outside the bounds, we can conclude that some type of AR process exists in the residuals. On the other hand, in figure 22, PACFs spike. That is, only the first and second bars (PACFs at lags 1 and 2) cross the bounds, indicating that no MA process is present in the residuals.

## C. Tests for an IID Sequence

Along with visual inspection, several statistics can be used to test whether residuals are observed values of an IID sequence. Among frequently employed tests are the portmanteau (Box-Pierce), Ljung-Box, McLeod-Li, turning point, difference-sign, rank, and runs tests. With these statistics, we test the following null hypothesis:

$H_0$: Residuals are a realization of an IID sequence.
$H_a$: Dependency exists among residuals.

The portmanteau, Ljung-Box, and McLeod-Li test statistics are compared to the chi-squared distribution. The turning point, difference-sign, rank, and runs test statistics are compared to the standardized normal distribution.

The purpose of these tests is different from the usual hypothesis tests, in which we determine whether to reject the null hypothesis in favor of the

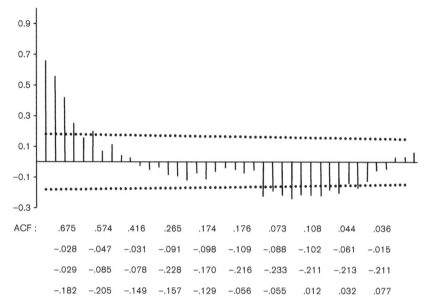

ACF :  .675  .574  .416  .265  .174  .176  .073  .108  .044  .036

      −.028  −.047  −.031  −.091  −.098  −.109  −.088  −.102  −.061  −.015

      −.029  −.085  −.078  −.228  −.170  −.216  −.233  −.211  −.213  −.211

      −.182  −.205  −.149  −.157  −.129  −.056  −.055  .012  .032  .077

FIGURE 21. Sample autocorrelation function (ACF) of the deseasonalized and detrended monthly violent crime rates, lags 1–40.

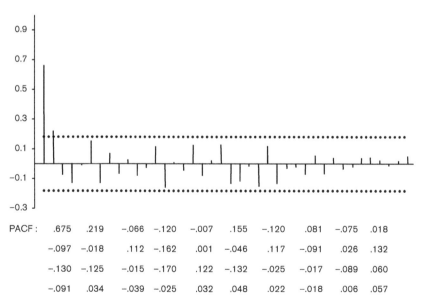

PACF :  .675  .219  −.066  −.120  −.007  .155  −.120  .081  −.075  .018

       −.097  −.018  .112  −.162  .001  −.046  .117  −.091  .026  .132

       −.130  −.125  −.015  −.170  .122  −.132  −.025  −.017  −.089  .060

       −.091  .034  −.039  −.025  .032  .048  .022  −.018  .006  .057

FIGURE 22. Sample partial autocorrelation function (PACF) of the deseasonalized and detrended monthly violent crime rates, lags 1–40.

research hypothesis. For example, in multiple OLS regression analysis, we attempt to show that at least some of our estimated regression coefficients are significantly different from zero by rejecting the null hypothesis that all of them are zero. In the following randomness tests, we attempt to fail to reject the null hypothesis that no systematic patterns exist in the residuals. It becomes more difficult to reject the null hypothesis as a level of significance— a probability threshold below which the null hypothesis will be rejected— becomes smaller. Therefore, it becomes easier to pass the randomness tests, as a level of significance becomes smaller. For example, it will be easier to fail to reject the null hypothesis and conclude that the coefficients of an AR($p$) process are not significant, when a level of significance is .01 than when it is .05. For the level of significance, .05 is commonly used.

*The portmanteau, Ljung-Box, and McLeod-Li Tests.* The portmanteau test utilizes the characteristic that the sum of squares of sample autocorrelations is approximated by the chi-squared distribution with $h$ degrees of freedom, if residuals are a realization of an IID or white sequence (Box and Pierce 1970). The test statistic is

$$Q = n \sum_{t=1}^{h} \rho_{y(t)}^{2}$$

where $n$ is the number of observations, $h$ is a lag value, $\rho_{y(t)}$ is the sample autocorrelation for a residual at lag $t$, and $t = 1, 2, \ldots, h$.

With this test statistic, we test whether all the autocorrelations from lag 1 to lag $h$ are not significantly different from zero (the null hypothesis) or some among them have nonzero values (the alternative hypothesis). We compare $Q$ to $\chi_{1-\alpha}^{2}(h)$, the $1 - \alpha$ percentile of the chi-squared distribution with $h$ degrees of freedom, where $\alpha$ is usually .05. If $Q$ is greater than the chi-squared value, we reject the null hypothesis at level $\alpha$.

The Ljung-Box and McLeod-Li tests are refinements of the portmanteau test (Ljung and Box 1978; McLeod and Li 1983). They are better approximated by $\chi_{1-\alpha}^{2}(h)$. The Ljung-Box test statistic is

$$Q_{LB} = n(n+2) \sum_{t=1}^{h} \left\{ \rho_{y(t)}^{2} / (n-t) \right\}$$

The McLeod-Li test statistic is

$$Q_{ML} = n(n+2)\sum_{t=1}^{b}\left\{\rho_{yy(t)}^2/(n-t)\right\}$$

The sample autocorrelations $\rho_{y(t)}$ of the residuals, $\{Y_t\}$, in the Ljung-Box test are replaced by the sample autocorrelations $\rho_{yy(t)}$ of the squared residuals, $\{Y_t^2\}$, in the McLeod-Li test. Of the three tests, the McLeod-Li test is considered as best approximated by $\chi_{1-\alpha}^2(b)$ (McLeod and Li 1983). As in the portmanteau test, if $Q_{LB}$ or $Q_{ML}$ is greater than $\chi_{1-\alpha}^2(b)$, we reject the null hypothesis at level $\alpha$.

*The Turning Point Test.* The turning point test checks whether residuals fluctuate more rapidly or more slowly than expected for an IID sequence. When we look at every three consecutive observations, $Y_{t-1}$, $Y_t$, and $Y_{t+1}$, the current observation, $Y_t$, is a turning point in the following six cases:

[1] $Y_{t-1} < Y_t > Y_{t+1}$

[2] $Y_{t-1} > Y_t < Y_{t+1}$

[3] $Y_{t-1} < Y_t = Y_{t+1}$

[4] $Y_{t-1} > Y_t = Y_{t+1}$

[5] $Y_{t-1} = Y_t > Y_{t+1}$

[6] $Y_{t-1} = Y_t < Y_{t+1}$

The current observation, $Y_t$, is not a turning point in the remaining three cases:

[7] $Y_{t-1} < Y_t < Y_{t+1}$

[8] $Y_{t-1} > Y_t > Y_{t+1}$

[9] $Y_{t-1} = Y_t = Y_{t+1}$

Thus, the probability that an observation will be a turning point when a time series is a realization of an IID sequence is ⅔.

The number of turning points can be counted by referring to residual plots. Sometimes, however, the difference between two adjacent points is very subtle, and we may not clearly detect the difference in residual plots. Thus, it is recommended to count the number of turning points by referring to residual values. In the deseasonalized and detrended monthly violent crime rates, for example, there are 66 turning points (figure 17).

Define $T$ to be the number of turning points. The expected number of turning points of an IID sequence of length $n$ is

$$E(T) = \frac{2(n-2)}{3}$$

since there will be $n - 2$ possible turning points (the first and last residuals cannot be turning points). The expected number of turning points in the deseasonalized and detrended monthly violent crime rates is 78.[12]

The variance of turning points for an IID sequence is

$$V(T) = \frac{(16n - 29)}{90}$$

The large positive value of $T - E(T)$ indicates that residuals tend to fluctuate more rapidly than expected for an IID sequence. In this case, there is a negative correlation between neighboring observations. On the other hand, the large negative value of $T - E(T)$ indicates that there is a positive correlation between neighboring observations. If $n$ is large, $T$ is well approximated by a normal distribution, $N(E(T), V(T))$. We standardize $T$ to utilize the standardized normal distribution, $N(0, 1)$:

$$Z = \frac{|T - E(T)|}{\sqrt{V(T)}}$$

We compare this statistic to the $1 - \alpha/2$ percentile of standardized normal distribution. For $\alpha$, .05 is commonly used, and the corresponding value is 1.96. If $Z$ is greater than the $1 - \alpha/2$ percentile, we reject the null hypothesis at level $\alpha$.

*The Difference-Sign Test.* This test checks whether there is dependency among residuals especially in the form of an increasing (or decreasing) trend. Define $S$ to be the number of times $Y_t - Y_{t-1}$ is positive. In the deseasonalized and detrended monthly violent crime rates (figure 17), for example, $S$ is 57. If the residuals are observed values of an IID sequence, the expected value of $S$ is

$$E(S) = \frac{(n-1)}{2}$$

since the probability that $Y_t - Y_{t-1} > 0$ will be ½, and $Y_t - Y_{t-1}$ will leave $n - 1$ cases. The variance of $S$ is

$$V(S) = \frac{(n+1)}{12}$$

For large $n$, $S$ will be approximated by a normal distribution with mean $E(S)$ and variance $V(S)$. We standardize $S$ to compare it to the standardized normal distribution:

$$D = \frac{|S - E(S)|}{\sqrt{V(S)}}$$

If $D$ is greater than the $1 - \alpha/2$ percentile of standardized normal distribution, we reject the null hypothesis at level $\alpha$.

With the difference-sign test, we can also know the type of dependency among residuals. When a perfect decreasing linear trend is present in residuals, $S$ will be 0. When the opposite is the case, $S$ will be $n - 1$. A large positive (negative) value of $S - E(S)$ indicates the presence of an increasing (decreasing) trend. If $S - E(S)$ nears zero, no upward or downward trends will be clearly discernible.

When seasonal components or smoothness are present, the difference-sign test may be inaccurate. Each cyclical period will consist of roughly half observations with an increasing trend and the other half observations with a decreasing trend. Since we have removed the seasonal components by the time we check the randomness of residuals, we may not need to worry about this possibility. However, if residuals show smoothness, values of residuals repeat increase and decrease, although irregularly. Consequently, the number of times that $Y_t - Y_{t-1}$ is positive may be closer to $E(S)$, compared with when residuals do not show smoothness.

For example, smoothness is apparent in figure 17, and $S$ is very close to $E(S)$, 59.5. The smoothness of residual plots indicates that some type of dependency exists in the residuals. However, as $S$ is closer to $E(S)$, the chance that we fail to reject the null hypothesis that the residuals are a realization of an IID sequence increases. Thus, when residuals show smoothness, the difference-sign test is less efficient, and we had better take the result with caution.

*The Rank Test.* The rank test is a refinement of the difference-sign test. The difference-sign test compares every two adjacent residuals, $Y_i$ and $Y_{i-1}$.

The rank test compares all the pairs $\{Y_i, Y_j\}$ such that $Y_j > Y_i$, where $j > i, i = 1, 2, \ldots, n - 1$, and $j = 2, 3, \ldots, n$.

Define $P$ to be the total number of these pairs. In the deseasonalized and detrended monthly violent crime rates, $P$ is 3,550. We should note that we compare the current observations with past observations both in the difference-sign test and in the rank test. We usually align our observations from the earliest observation to the latest observation. Thus, we begin counting $S$ and $P$ with the last case, for example the 120th case in the monthly violent crime rates. If we align our observations from the latest observation to the earliest observation, we begin with the first case. $P$ is $n(n - 1)/2$ when a perfect increasing linear trend is present and zero when a perfect decreasing linear trend is present. If a perfect increasing linear trend is present in the deseasonalized and detrended monthly violent crime rates, there will be 7,140 pairs. If the residuals are observed values of an IID sequence, the mean of $P$ is

$$E(P) = \frac{n(n-1)}{2} \times \frac{1}{2} = \frac{n(n-1)}{4}$$

since the probability that $Y_j > Y_i$ will be ½. The variance of $P$ is

$$V(P) = \frac{n(n-1)(2n+5)}{72}$$

For large $n$, $P$ will be approximated by a normal distribution with $E(P)$ as mean and $V(P)$ as variance. Thus, the standardized value of $P$ can be used to test the randomness of residuals:

$$R = \frac{|P - E(P)|}{\sqrt{V(P)}}$$

If $R$ is greater than the $1 - \alpha\!/_2$ percentile of standardized normal distribution, we reject the null hypothesis. Since the rank test considers all the possible pairs in $\{Y_i\}$ such that $Y_j > Y_i$, the effect of cyclical components or smoothness on the test statistic is less serious than in the difference-sign test. When cyclical components or smoothness are obvious, however, we still had better take the result with caution. Like the difference-sign test, the large positive (negative) value of $P - E(P)$ indicates the presence of an increasing (decreasing) trend in residuals. If $P - E(P)$ is close to zero, no clearly discernible trend is present in the residuals.

*The Runs Test.* This test was originally invented to check whether two separate samples are from the same population (Blalock 1979). If two samples are from the same population, they will behave very similarly, and thus they will be mixed well. A *run* is a sequence of values from the same sample, when two samples are mixed and ranked. If two samples are from different populations, they will not be mixed well, and thus the number of runs will be significantly smaller than expected.

We can use the runs test to check the randomness of residuals. In this case, the runs test checks whether the order of occurrence of two values of a variable is random. To identify a run, we first need to specify a cut point so that we can dichotomize a time series into one group of cases with values less than the cut point and the other with values greater than or equal to the cut point. This is the same as creating two separate samples to check whether their behaviors show significantly different patterns. A sequence of values above or below a specified cut point consists of a run.

The cut point that is usually selected is the median, mean, or mode. Any of these three is appropriate as long as the distribution of residuals is close to the normal distribution. We will have roughly the same number of cases in the two groups. If the distribution is highly skewed or if there are extreme cases, the median is more appropriate as a cut point than the others.

In figure 17, for example, when we select the mean as a cut point, there are 28 runs:

$$+ (5), - (2), + (8), - (6), + (1), - (1), + (1), - (2), + (1), - (9), + (9), - (1), + (1), - (2), + (2), - (8), + (1), - (12), + (1), - (8), + (2), - (1), + (1), - (2), + (10), - (3), + (13), - (7)$$

where + indicates cases whose values are greater than or equal to the mean, − indicates cases whose values are smaller than the mean, and the number in the parentheses is the number of elements of each run.

Define $R$ to be the number of runs. For large $n$, the sampling distribution of $R$ is approximately normal with mean:

$$E(R) = \frac{(n_1 + n_2)(n_1 + n_2 + 1) - (n_1^2 + n_2^2)}{n_1 + n_2} = \frac{2n_1 n_2}{n_1 + n_2} + 1$$

where $n_1$ is the number of cases whose values are greater than or equal to the cut point and $n_2$ is the number of cases whose values are smaller than the cut point.

In the deseasonalized and detrended monthly violent crime rates, $n_1$ is 56 and $n_2$ is 64. The variance is

$$V(R) = \frac{\left[(n_1^2 + n_2^2)\left\{(n_1^2 + n_2^2) + (n_1 + n_2)(n_1 + n_2 + 1)\right\} - 2(n_1 + n_2)(n_1^3 + n_2^3) - (n_1 + n_2)^3\right]}{(n_1 + n_2)^2(n_1 + n_2 - 1)}$$

$$= \frac{2n_1 n_2 (2n_1 n_2 - n_1 - n_2)}{(n_1 + n_2)^2(n_1 + n_2 - 1)}$$

In order to utilize the standardized normal distribution, we standardize $R$:

$$Z = \frac{|R - E(R)|}{\sqrt{V(R)}}$$

$R$ around its expected value indicates that no significant difference exists between the two groups that were separated by the cut point. In other words, cases of the two groups are mixed well enough to be considered as representing the same population. In a time series, this will indicate that the residuals as a whole are observed values of an IID sequence. The smoothness of residuals, if any, will be negligible, and we can consider that their fluctuations are only random.

On the other hand, too many or too few runs will indicate the presence of a systematic pattern in the residuals, that is, negative or positive serial correlation between neighboring observations. A number of runs smaller than expected indicate the smoothness of residual plots, that is long stretches of plots on the same side above or below the cut point, and the presence of positive correlation between neighboring residuals. A number of runs bigger than expected indicate the presence of negative correlation between residuals. Thus, if $Z$ is greater than the $1 - \alpha/2$ percentile of the standardized normal distribution, we reject the null hypothesis at level $\alpha$.

In the original runs test, which checks whether two samples are from the same population, too many runs is not a problem because it indicates that two samples are from the same population. Perhaps for this reason, caring primarily about too few runs, some statistical programs, such as Minitab and SPSS, do not take the absolute value of $Z$. Thus, if $R$ is smaller than $E(R)$, these programs will report a negative test statistic. In this case, we reject the null hypothesis if $Z$ is smaller than the left-hand $(1 - \alpha/2)$ percentile of the standardized normal distribution, for example, $-1.96$ at $\alpha = .05$.

*The Durbin-Watson d Test.* The Durbin-Watson $d$ test statistic is the ratio of the sum of the squared deviation between current residuals and lag-1 previous residuals to the sum of the squared current residuals:

$$\frac{\sum_{t=2}^{n}(Y_t - Y_{t-1})^2}{\sum_{t=1}^{n}Y_t^2}$$

where $n$ is the number of cases included in the analysis.

The Durbin-Watson $d$ test statistic is similar to the first-order autocorrelation in residuals, that is with the bivariate correlation between $\{Y_t\}$ and $\{Y_{t-1}\}$:

$$r = \frac{\sum_{t=2}^{n}(Y_t - \overline{Y_t})(Y_{t-1} - \overline{Y_{t-1}})}{\sqrt{\sum_{t=2}^{n}Y_t^2 \sum_{t=2}^{n}Y_{t-1}^2}} = \frac{\sum_{t=2}^{n}(Y_t \times Y_{t-1})}{\sqrt{\sum_{t=2}^{n}Y_t^2 \sum_{t=2}^{n}Y_{t-1}^2}}$$

In this equation $E(Y_t)$ is zero and $\sum_{t=1}^{n}Y_t^2$, $\sum_{t=2}^{n}Y_t^2$, and $\sum_{t=2}^{n}Y_{t-1}^2$ are almost identical. The difference between $\sum_{t=1}^{n}Y_t^2$ and $\sum_{t=2}^{n}Y_t^2$ is the squared residual that is associated with the first observation, $Y_1^2$. The difference between $\sum_{t=1}^{n}Y_t^2$ and $\sum_{t=2}^{n}Y_{t-1}^2$ is the squared residual that is associated with the last observation, $Y_n^2$.

Therefore, the Durbin-Watson $d$ test checks whether significant first-order serial correlation is present in the residuals.[13] Statistical programs usually include the Durbin-Watson $d$ statistic as an option in the linear regression estimation procedure. Some stand-alone time series programs, such as EViews, automatically calculate this statistic.

When there is no first-order autocorrelation in residuals, that is when $\sum_{t=2}^{n}(Y_t \times Y_{t-1})/\sum_{t=1}^{n}Y_t^2$ is zero, the Durbin-Watson $d$ test statistic will be approximately 2, because the numerator will be about twice the denominator:

$$\frac{\sum_{t=2}^{n}(Y_t - Y_{t-1})^2}{\sum_{t=1}^{n}Y_t^2} = \frac{\sum_{t=2}^{n}(Y_t^2 - 2Y_t \times Y_{t-1} + Y_{t-1}^2)}{\sum_{t=1}^{n}Y_t^2}$$

$$= \frac{-2\sum_{t=2}^{n}(Y_t \times Y_{t-1})}{\sum_{t=1}^{n}Y_t^2} + \frac{\sum_{t=2}^{n}(Y_t^2 + Y_{t-1}^2)}{\sum_{t=1}^{n}Y_t^2} \approx \frac{2\sum_{t=2}^{n}Y_t^2}{\sum_{t=1}^{n}Y_t^2}$$

The Durbin-Watson $d$ statistic ranges from 0 to 4. If there is a positive correlation $(0 < r \leq 1)$ between $Y_t$ and $Y_{t-1}$, $-2\left\{\sum_{t=2}^{n}(Y_t \times Y_{t-1}) \div \sum_{t=1}^{n} Y_t^2\right\}$ will be greater than or equal to $-2$ and smaller than 0, and therefore the Durbin-Watson $d$ test statistic will be greater than or equal to 0 and smaller than 2. If there is a negative correlation $(-1 \leq r < 0)$ between $Y_t$ and $Y_{t-1}$, $-2\left\{\sum_{t=2}^{n}(Y_t \times Y_{t-1}) \div \sum_{t=1}^{n} Y_t^2\right\}$ will be greater than 0 and smaller than or equal to 2, and therefore the Durbin-Watson $d$ test statistic will be greater than 2 and smaller than or equal to 4. Usually, with $d$ smaller than 1 or greater than 3, we worry about the possibility that a serious positive or negative first-order autocorrelation is present in the residuals.

The actual levels of significance of the Durbin-Watson $d$ test statistic vary for the sample size and the number of independent variables, including the constant. The upper and lower bounds of the confidence interval for $d$ can be determined by referring to the $d$ statistic table (table A1, appendix 4, page 205). In the table, if we do not find a row for the sample size, we go to the next-lowest sample size. For example, if the number of cases is 47, we go to the row with the sample size equal to 45. With the $d$ statistic, we test the following null hypothesis:

$H_0$: There is no first-order autocorrelation in the residuals.
$H_a$: The current residuals and the lag-1 previous residuals are positively (or negatively) correlated.

If the $d$ statistic is smaller than 2, we test the null hypothesis against the alternative hypothesis of positive first-order autocorrelation in residuals. If $d$ is smaller than the lower bound $d_L$, we reject the null hypothesis, and if it is greater than the upper bound $d_U$, we fail to reject the null hypothesis. If $d$ is between the lower bound and the upper bound, the test is inconclusive.

If $d$ is greater than 2, we test the null hypothesis against the alternative hypothesis of negative first-order autocorrelation in residuals. In this case, if $4 - d$ is smaller than the lower bound, we reject the null hypothesis, and if it is greater than the upper bound, we fail to reject the null hypothesis. If $4 - d$ is between the lower bound and the upper bound, the test is inconclusive.

The Durbin-Watson $d$ statistic is applicable only to first-order, and not to higher-order, autocorrelations (Johnston 1984). In addition, the Durbin-

Watson $d$ statistic should not be used if any independent variables are correlated with residuals. In this case, the Durbin-Watson $d$ statistic can be biased and we may fail to reject the null hypothesis when it is actually false. In multiple time series analysis, for example, the Durbin-Watson $d$ statistic should not be used when lagged dependent variables are included as independent variables. Any lagged dependent variable on the right-hand side of the equation can be correlated with residuals, which are the differences between the observed and estimated values of the dependent variable. If we include a lagged dependent variable as an independent variable, we can employ Durbin's alternative tests, such as the Durbin's $h$ test (see chapter 6).

All these tests check whether residuals are a realization of an IID sequence. As mentioned above, we can determine whether residuals are a realization of an IID sequence by visually inspecting the scatter plots and sample ACF and PACF correlograms of residuals. Statistical tests are only of complementary use. We need to refer to statistical tests to confirm what we find from our visual inspection of residuals, rather than to determine whether residuals are a realization of an IID sequence without visually inspecting residuals.

In addition, to be more confident about test results, we had better employ several tests together, rather than relying on just one of them. Some tests may fail to reject the null hypothesis, while other tests reject it. We need to check whether tests that reject the null hypothesis outnumber tests that fail to reject it.

Statistical programs provide different test statistics. No matter what programs we use, we can calculate all the test statistics explained above. For the portmanteau, Ljung-Box, and McLeod-Li tests, we can use sample autocorrelations that are reported by statistical programs. For the McLeod-Li test, we first square residuals, and then we let statistical programs report sample autocorrelations of squared residuals. These test statistics are usually calculated with the first 24 cases for monthly time series and the first 8 cases for quarterly time series. In most cases, test statistics are calculated with less than one-fourth of all the cases because autocorrelations can be accurately estimated with less than one-fourth of all the cases, if the total number of cases is greater than 50 (Box and Jenkins 1976).

It will be easier to calculate the turning point, difference-sign, rank, and runs test statistics. But it may take time to count $T$, $S$, $P$, and $R$ because we need to calculate these statistics with all the cases.

TABLE 3. Randomness Test Statistics for the Deseasonalized and Detrended
Monthly Violent Crime Rates, $\{Y_t\}$

| Test | Observed value | Critical value at $\alpha$ = .05 | IID hypothesis |
| --- | --- | --- | --- |
| Ljung-Box | 155.79 | 31.4 | Rejected |
| Mcleod-Li | 52.77 | 31.4 | Rejected |
| Turning-point | 2.62 | 1.96 | Rejected |
| Difference-sign | 0.79 | 1.96 | Not rejected |
| Rank | 0.09 | 1.96 | Not rejected |
| Runs | −6.03 | −1.96 | Rejected |

NOTE. The Ljung-Box and Mcleod-Li tests were calculated with the residuals from lag 1 to lag 24. The other tests were calculated with all 120 residuals.

## D. The Case of Monthly Violent Crime Rates, 1983–1992

Table 3 presents test statistics for the randomness of the deseasonalized and detrended crime rates. Through visual inspection, we have already detected that some type of systematic patterns exist in the deseasonalized and detrended crime rates. In table 3, four of the six test statistics detect significant deviation from an IID sequence, making us reject the IID hypothesis. The difference-sign and rank tests fail to reject the null hypothesis that the deseasonalized and detrended crime rates are a realization of an IID sequence. As explained above, however, these two test statistics are influenced by the smoothness of the deseasonalized and detrended crime rates (figure 17).

In addition, the Durbin-Watson $d$ statistic indicates the presence of a serial correlation in the residuals. Since the number of cases is 120 and the number of independent variables including the constant is 2, the lower bound of the Durbin-Watson $d$ statistic is 1.634 at the level of significance, .05 (table A1). Since this lower bound is greater than 0.563, we reject the null hypothesis and conclude that a positive serial correlation is present in the residuals. Our next step is to find a probabilistic model that can represent the dependency among the deseasonalized and detrended crime rates.

## (5) FITTING THE RESIDUALS

Even after a time series is accurately deseasonalized and detrended, AR and/ or MA processes are often present in the residuals. To complete time series modeling, we need to estimate and then eliminate such processes from the residuals, leaving only white noise.

As explained above, we can easily detect the dependency of a deseasonalized and detrended time series by visually inspecting residual plots and sample ACF and PACF correlograms. To model the dependency, we first need to determine the orders of AR and/or MA processes. Based on the orders, we estimate the coefficient of an AR process, $\phi_i$, and/or the coefficient of an MA process, $\theta_j$. Then, by removing the estimated processes from the residuals, we obtain white (or IID) noise with a mean equal to zero and variance equal to $\sigma^2$.

## A. AR(p), MA(q), and ARMA(p, q) Models

Conventionally, $p$ denotes the order of AR process. It indicates how many previous values in a time series are significantly correlated with its current value and thus how many previous values need to be used to explain the current value. We estimate an AR($p$) model after we deseasonalize and/or detrend a time series. Therefore, the current or previous observed values are those of the residuals, not of the original observations.

A pure AR(1) model, for example, is

$$Y_t = \phi Y_{t-1} + Z_t$$

where $|\phi| < 1$.

In this model, the current observation, $\{Y_t\}$, is explained with the lag-1 previous observation, $\{Y_{t-1}\}$, and white noise, $\{Z_t\}$. There is a significant relationship between $\{Y_t\}$ and $\{Y_{t-1}\}$, but correlations between $\{Y_t\}$ and the lag-2 or more previous observations are not significantly different from zero.

If $|\phi| = 1$, all the previous observations will equally determine the current observation:

$$Y_t = Y_{t-1} + Y_{t-2} + Y_{t-3} + \ldots + Y_{t-p} + Z_t$$

A *unit root* ($|\phi| = 1$) that indicates no decaying memory or no diminishing effect is not theoretically plausible in the social sciences. Therefore, we usually consider a unit root as a problem to be solved to model a time series accurately in the social sciences. We can check for the presence of a unit root with the Dickey-Fuller test (see chapter 6, section 3B). If $|\phi|$ is not significantly smaller than 1, we need to difference a time series to remove the unit root before we estimate the ARMA($p$, $q$) model. Thus, we estimate an *autoregressive integrated moving average* (ARIMA) model, in which the I indicates that a time series is differenced. In this case, we estimate an ARIMA($p$, 1, $q$) model, where 1 is the order of differencing.[14]

An AR coefficient greater than 1 is called an *explosive AR coefficient.* Explosive AR coefficients are not realistic. When an AR coefficient is explosive, temporally more distant observations have greater influence over the current observation. For example, if $\phi = 2$ in an AR(1) process, a lag-$k$ previous observation will have $2^{k-1}$ times greater influence than the lag-1 previous observation over the current observation:

$$Y_t = 2Y_{t-1} + 4Y_{t-2} + 8Y_{t-3} + \ldots + 2^k Y_{t-k} + Z_t$$

In addition, an explosive AR process can be stationary only when current values are explained by future values:

$$Y_t = \phi Y_{t-1} + Z_t = \frac{1}{\phi} Y_{t+1} - \frac{1}{\phi} Z_{t+1}$$

The letter $q$ denotes the order of an MA process. It indicates how many previous white (or IID) noise components are used to explain the current value. A pure MA(1) model, for example, is

$$Y_t = Z_t + \theta Z_{t-1}$$

In this model, any previous observations are not significantly correlated with the current observation, but the lag-1 previous white noise component, $\{Z_{t-1}\}$, is significantly correlated with the current observation, $\{Y_t\}$. Thus, the current observation, $\{Y_t\}$, is explained with the current and lag-1 previous white noise components.

Sometimes, a time series can be defined more effectively with a combination of AR($p$) and MA($q$) processes. In an ARMA(1, 1) model, for example, the current residual, $\{Y_t\}$, is explained from its lag-1 previous observation, $\{Y_{t-1}\}$, the current white noise, $\{Z_t\}$, and the lag-1 previous white noise, $\{Z_{t-1}\}$:

$$Y_t = \phi Y_{t-1} + Z_t + \theta Z_{t-1}$$

*Determining* p *and* q. Fortunately, in most social science studies, it is quite simple to determine $p$ and $q$. They will be 0 or 1 or 2 in most cases. We can easily determine $p$ and $q$ by visually inspecting sample ACFs and sample

PACFs. *Autocorrelation* at $k$th lag is correlation between a time series that begins with the current (lag-0) observation and a part of the series that begins with the lag-$k$ observation:

$$\text{ACF}(k) = \frac{\sum_{i=1}^{n-k}(Y_i - \overline{Y})(Y_{i+k} - \overline{Y})}{\sum_{i=1}^{n}(Y_i - \overline{Y})^2}$$

ACF at lag 1, for example, is correlation between a time series that begins with the current (lag-0) observation and a part of the series that begins with the lag-1 observation. In a pure AR(1) process, only the lag-1 observation is significantly related to the current observation. Therefore, the AR coefficient $\phi$ is the ACF at lag 1. ACF(2) is $\phi^2$, ACF(3) is $\phi^3$, . . . , and ACF($k$) is $\phi^k$.[15]

For an AR($p$) process ($p > 1$), ACFs are not represented directly by the coefficients of the AR($p$) process. We can calculate ACFs via the Yule-Walker equations (Brockwell and Davis 2002). For example, the lag-1 and lag-2 observations are significantly related to the current observation in a pure AR(2) process:

$$Y_t = \phi_1 Y_{t-1} + \phi_2 Y_{t-2} + Z_t$$

In this case, the ACF at lag 1 is

$$\rho_1 = \phi_1 + \phi_2\,\rho_1 = \phi_1 / (1 - \phi_2)$$

The ACF at lag 2 is

$$\rho_2 = \phi_1\rho_1 + \phi_2 = \phi_1^2 / (1 - \phi_2) + \phi_2$$

Then the ACF at lag $k$, $\rho_k$, is

$$\rho_k = \phi_1\rho_{k-1} + \phi_2\,\rho_{k-2}$$

In an AR($p$) process, $Y_t = \phi_1 Y_{t-1} + \phi_2 Y_{t-2} + \ldots + \phi_p Y_{t-p} + Z_t$, the ACF at lag $k$ ($k \le \text{p}$) is

$$\rho_k = \phi_1\rho_{k-1} + \phi_2\rho_{k-2} + \ldots + \phi_{p-1}\rho_{k-p-1} + \phi_p\rho_{k-p} = \sum_{i=1}^{p}\phi_i\rho_{k-i}$$

In this equation, autocorrelation at lag $k$ is equal to autocorrelation at lag $-k$.

*Partial autocorrelation* at lag $k$ is correlation between a time series and its $k$th lag after the correlations at intermediate lags have been partialled out. This is why it is termed partial autocorrelation. Both ACF and PACF at lag 0 are 1 because they are correlations between the same time series. PACF at lag 1 is identical to ACF at lag 1 because there are no intermediate lags between lag 0 and lag 1. In PACF at lag 2, ACF at lag 1 has been partialled out from the ACF at lag 2:

$$PACF(2) = \frac{ACF(2) - ACF(1)^2}{1 - ACF(1)^2}$$

The calculation of PACF($k$) is equivalent to that of partial correlation. We calculate correlation between two time series that begin at lag 0 and lag $k$, respectively, while controlling for the effects of one or more time series that begin at lags from 1 to $k - 1$. PACF(2), for example, is correlation that remains between two time series that begin at lag 0 and lag 2, respectively, after we controlled for the correlation that is due to their mutual association with a time series that begins at lag 1.

When a time series is a realization of a white or IID sequence with a zero mean and a constant variance, both ACFs and PACFs at all lags will not be significantly different from zero. Only a few occasional ACFs or PACFs especially at higher lags may be significantly different from zero by chance alone. If any type of dependency exists in a time series, however, some ACFs and PACFs will be significantly different from zero, and thus they will fall outside the confidence bounds. As mentioned above, this behavior of ACFs and PACFs can be utilized to detect the dependency among residuals and to determine the orders of AR and/or MA processes.

In an AR($p$) process, ACFs will abruptly decay as lags increase because ACF at lag $k$ is $\phi^k$ for AR(1) process or $\sum_{i=1}^{p} \phi_i \rho_{k-i}$ for an AR($p$) process ($p > 1$) and $|\phi|$ should be smaller than 1. PACFs whose lags are greater than $p$ will not be significantly different from zero because all significant ACFs, ACF(1) to ACF($p$), are partialled out. Among PACFs beyond lag $p$, only a few at higher lags may fall outside the confidence bounds by chance alone. For this reason, when a pure AR($p$) process is present, PACFs will clearly spike $p$ times. Thus, with decaying ACFs, we can detect the presence of an AR($p$) process. Then, from the number of spikes in the PACFs, we can determine the order of the AR process. According to figures 21 and 22, for instance, we can determine that an AR(2) process is present in the deseasonalized and detrended monthly violent crime rates from 1983 to 1992.

When a time series contain seasonal components, we can expect additional spikes. For example, when an AR process with a seasonality of lag $k$ is present in a time series, PACF at lag $k$ (and maybe at lags $2k$ and $3k$) may spike. PACFs around lag $k$ may also spike (see e.g. figure 41, chapter 6, page 128). Therefore, we should not consider these spikes when we determine the order of an AR process. Instead, we need to return to the original time series and begin the modeling procedure again with deseasonalization.

If ACF at lag 1 is positive, ACFs of an AR(1) process will be positive and abruptly decay and PACFs will show a positive spike at lag 1. If ACF at lag 1 is negative, meaning that the current observation is negatively correlated with the lag-1 previous observation, ACFs of AR(1) process will begin with a negative spike at lag 1 and show an abrupt oscillating decay of positive and negative spikes. In this case, PACFs will show a negative spike at lag 1.

ACF at lag 2 of an AR(2) process will be positive when ACF at lag 1 $(\phi_1)$ is positive. ACF at lag 2 will be positive in most cases when ACF at lag 1 $(\phi_1)$ is negative, because $\phi_1$ is squared for ACF at lag 2. When ACFs at lags 1 and 2 are positive, ACFs of the AR(2) process will abruptly decay and PACFs will show positive spikes at lags 1 and 2. When ACF at lag 1 is negative and ACF at lag 2 is positive, ACFs of the AR(2) process will show an abrupt oscillating decay and PACFs will show a negative spike at lag 1 and a positive spike at lag 2. Figure 23 shows the model ACF and PACF correlograms of an AR($p$) process.

When a pure MA($q$) process is present, PACFs will decay as lags increase.[16] On the other hand, ACFs will spike $q$ times, because, through the lag-1 to lag-$q$ previous white noise components, $\{Z_{t-1}, \ldots , Z_{t-q}\}$, the current observation, $\{Y_t\}$, will be related to the lag-1 to lag-$q$ previous observations, $\{Y_{t-1}, \ldots , Y_{t-q}\}$.[17] Thus, when a pure MA($q$) process is present, we should observe PACFs, not ACFs, similar to those in figure 21. In this case, we can count the number of spikes in ACFs to determine the order of the MA process. If ACFs look like figure 22, for instance, an MA(2) process is present in the residuals.

In a pure MA(1) process, if $\theta$ is positive, ACF will show a negative spike at lag 1 and PACFs will be negative and abruptly decay. If $\theta$ is negative, ACF will show a positive spike at lag 1 and PACFs will show an abrupt oscillating decay.

In a pure MA(2) process, if $\theta_1$ and $\theta_2$ are positive, ACF will show negative spikes at lags 1 and 2, and PACFs will show an abrupt decay of negative spikes. If $\theta_1$ and $\theta_2$ are negative, ACF will show positive spikes at lags 1 and 2, and PACFs will show an abrupt oscillating decay. If $\theta_1$ is positive and $\theta_2$ is negative, ACF will show a negative spike at lag 1 and a positive spike at lag 2.

AR(1) process, $\phi > 0$

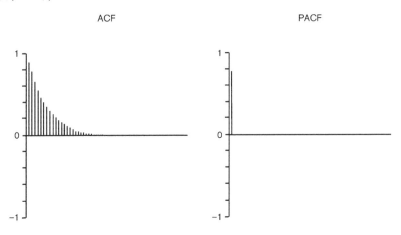

ACF                    PACF

AR(1) process, $\phi < 0$

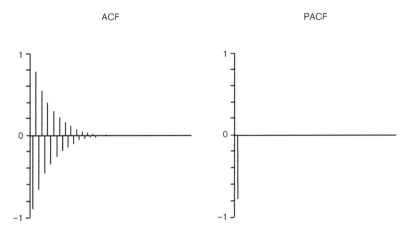

ACF                    PACF

FIGURE 23. Model autocorrelation function (ACF) and partial autocorrelation function (PACF) of AR(1) and AR(2) processes.

If $\theta_1$ is negative and $\theta_2$ is positive, ACF will show a positive spike at lag 1 and a negative spike at lag 2. PACFs will show an abrupt oscillating decay. Figure 24 shows the model ACF and PACF correlograms of an MA($q$) process.

When both AR($p$) and MA($q$) processes are present, it is difficult to determine $p$ and $q$ by visually inspecting ACF and PACF correlograms. We can try several models with different $p$ and $q$ values, checking selection criteria that are explained below. In most social science time series, however, AR($p$)

AR(2) process, $\phi_1 > 0$ and $\phi_2 > 0$

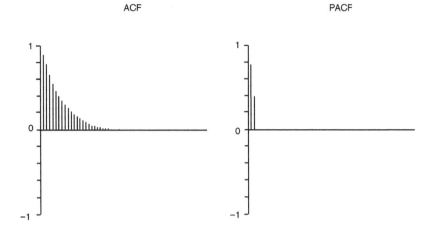

AR(2) process, $\phi_1 < 0$ and $\phi_2 > 0$

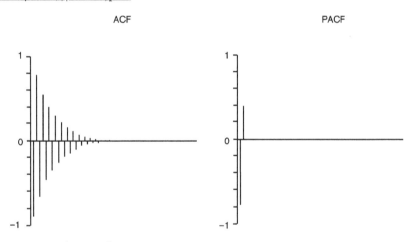

FIGURE 23. *(continued)*

and MA($q$) processes rarely exist together. When both processes are present, they will be represented by an ARMA(1, 1) process in most cases (McDowall et al. 1980). When an ARMA(1, 1) process is present and if both $\phi$ and $\theta$ are positive, both ACFs and PACFs will be positive and decline exponentially from the first lag. If both $\phi$ and $\theta$ are negative, both ACFs and PACFs will show an oscillating decay. If $\phi$ is positive and $\theta$ is negative, ACFs will be positive and abruptly decay and PACFs will show an oscillating decay. If $\phi$ is

MA(1) process, $\theta > 0$

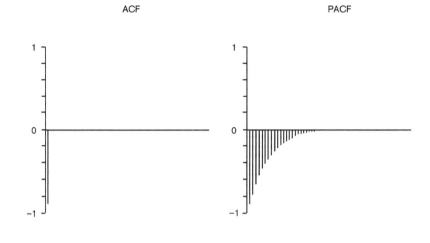

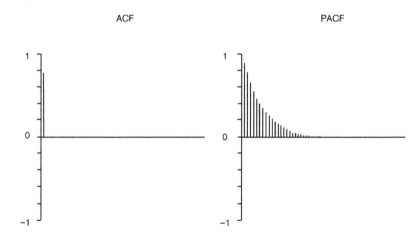

FIGURE 24. Model ACF and PACF of MA(1) and MA(2) processes.

negative and $\theta$ is positive, ACFs will show an oscillating decay and PACFs will be negative and abruptly decay. Figure 25 shows the model ACF and PACF correlograms of an ARMA(1, 1) process.

We cannot determine the true values of $p$ and $q$ of an ARMA($p$, $q$) model by visually inspecting ACF and PACF correlograms. Thus, when selecting an ARMA($p$, $q$) model based on our inspection of ACF and PACF correlograms and estimating its coefficients, we had better try competing models.

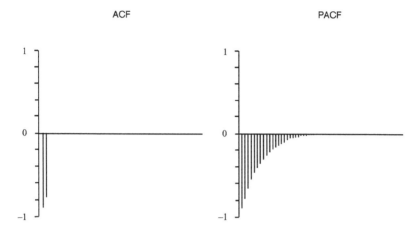

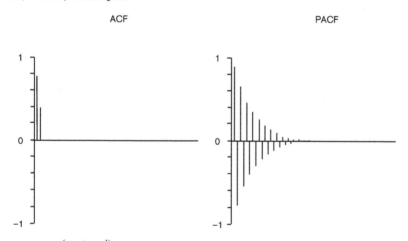

FIGURE 24. *(continued)*

In most social science studies, we need to try only a few competing models because the very high order of ARMA($p$, $q$) model is unlikely to exist. In most cases, the $p$ and $q$ of the competing models will be 0, 1, or 2. We can compare the selection criteria of the models (see below). Then, we can check the appropriateness of the selected model by checking the randomness of the residuals from which the selected process has been removed (see section 4C).

ARMA(1, 1) process, $\phi > 0$ and $\theta > 0$

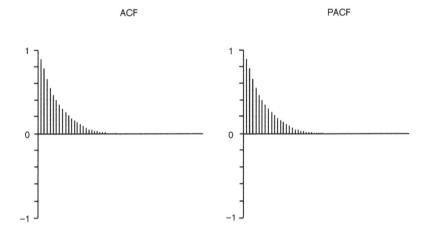

ARMA(1, 1) process, $\phi < 0$ and $\theta < 0$

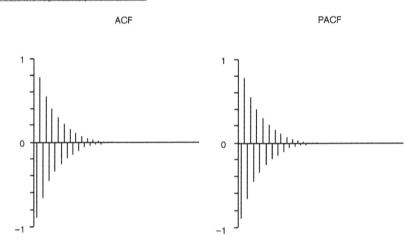

FIGURE 25. Model ACF and PACF of ARMA$(p, q)$ process.

*Optimization Process.* After we have determined $p$ and/or $q$, we begin the optimization process that searches for the maximum likelihood estimates of the coefficients, $\phi_i$ and $\theta_j$, of the ARMA$(p, q)$ model. Statistical programs first estimate some preliminary values for $\phi_i$ and $\theta_j$ with the given orders of the AR and/or MA processes. In the preliminary estimation process, the maximum likelihood estimation method is not employed. Instead, programs estimate the coefficients with several types of algorithms, such as the Burg,

ARMA(1, 1) process, $\phi > 0$ and $\theta < 0$

ACF                                    PACF

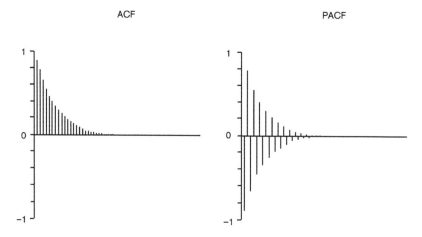

ARMA(1, 1) process, $\phi < 0$ and $\theta > 0$

ACF                                    PACF

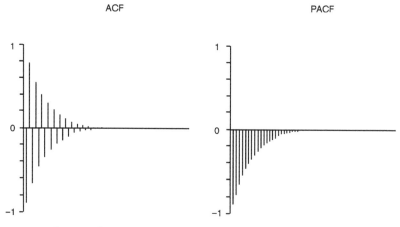

FIGURE 25. *(continued)*

Yule-Walker, Hannan-Rissanen, innovation, Melard's, and Kalman filtering algorithms.[18] We usually do not compare several competing models at the preliminary estimation stage. Therefore, most programs do not report the selection criteria for the preliminary estimates.

We may be asked to input preliminary values for $\phi_i$ and $\theta_j$. When programs do not automatically estimate the preliminary coefficients, we can determine them based on our prior knowledge. Then we can let statistical programs take the preliminary estimate as the starting point of the optimization process.

If preliminary estimates are bad, it may take longer for the optimization process to reach the maximum likelihood point, or it may fail to converge with the given number of iterations. Then we simply need to increase the number of iterations. In fitting residuals with the maximum likelihood estimation method, what we really need to care about is the orders of the AR and/or MA processes. If we select wrong orders of AR and/or MA processes, the estimated time series model will be inaccurate.

Beginning with the preliminary estimates, statistical programs numerically search for $\phi_i$ and $\theta_j$ whose likelihood is at its maximum. We do not know the true maximum likelihood parameters, and therefore we cannot know for sure whether we complete the optimization process. The best we can do is to predetermine the level of likelihood increase. Statistical programs determine that the estimate is close enough to the true maximum likelihood parameter, if the likelihood increase is smaller than a predetermined level, usually 0.001.

*Model Selection Criteria*

[1] RESIDUAL VARIANCE

Residual variance is a good indicator of the accuracy of an estimated time series model, as in ordinary least squares regression analysis. Among several competing models, a model that has the smallest residual variance is generally better than the others. There are two ways in which we can reduce residual variance. One is to increase the order, $p$ and/or $q$, of an ARMA process. As more parameter estimates are included in an ARMA$(p, q)$ model, residual variance is likely to decrease. As mentioned above, however, a high-order ARMA$(p, q)$ model is unreal in most social science time series.

The other way is to increase the number of observations. As the number of cases increases, statistical estimates will become closer to the true parameters. A drawback of this approach is that it is very difficult to collect more observations of a time series. We can wait for more observations, but the number of observations will not significantly increase in the near future. We may go back to the past, but it is often inappropriate to base our estimation on the observations that were collected long ago. Everything, such as sociodemographic composition of a society, cultures, and major political ideas, keeps changing in human society. We may not accurately explain current phenomena or forecast future outcomes if we rely on observations that were collected long ago.

Therefore, the best we can do to obtain a more accurate time series model is to try several competing models that have $p$ and/or $q$ less than or equal to 2, and select the one with the smallest residual variance.

The log-likelihood is the natural logarithm of the residual variance of the estimated AR($p$) model. The log-likelihood is negative, and a model with a greater log-likelihood (closer to zero) is generally better than the others. Some statistical programs report the log-likelihood value multiplied by –2. In this case, a model with a smaller log-likelihood (positive and closer to zero) is generally better.

Rather than comparing the log-likelihood, we can utilize the characteristic that if the residuals are normally distributed, the –2 log-likelihood ratio statistic

$$-2LLR = -2\ln(L_p/L_{p+k}) = -2\{\ln(L_p) - \ln(L_{p+k})\}$$

(where ln is natural logarithm and $L_p$ and $L_{p+k}$ are the log-likelihoods of AR($p$) and AR($p + k$) processes, respectively) is well approximated by a chi-squared distribution with $k$ degrees of freedom. We can compare the –2 log-likelihood ratio statistic of two competing pure AR processes. For example, we can compare AR($p$) and AR($p + k$) processes. If the statistic is greater than the chi-squared value with $k$ degrees of freedom at a given significance level (e.g. .05), we reject the null hypothesis that $k$ more parameter estimates do not significantly contribute to the explanation of a systematic pattern of residuals. We conclude that an AR($p + k$) process represents the systematic pattern of residuals better than does an AR($p$) process.

We can also test whether all AR coefficients are zero with the –2 log-likelihood ratio between a specified AR model with $k$ parameters, AR($k$), and a model with intercept alone ($p = 0$) (Aldrich and Nelson 1984; Menard 1995). In time series analysis, however, we can determine whether a systematic pattern exists in residuals by visually inspecting residuals and their ACF and PACF correlograms. Thus, we usually do not need to test whether the estimated coefficients of an AR($p$) process are not zero.

When using the residual variance or the log-likelihood as a criterion to select an AR($p$) model, we may be tempted to increase the order of our model because the residual variance will become smaller as more parameters are included in our model. In final prediction error, penalty factors are included to compensate for the decrease of residual variance due to the increase of the order of the AR($p$) model (Akaike 1969):

$$FPE = \frac{\sigma_M^2 (n + p)}{n - p}$$

where $\sigma_M^2$ is the maximum likelihood estimate of the residual variance of an AR($p$) process, $n$ is the number of observations, and $p$ is the order of the AR process.

As more parameters are included in the model, the residual variance $\sigma_M^2$ will steadily decrease, but $(n + p) / (n - p)$ will become greater at an increasing rate. If there are 100 observations, for example, $(n + p) / (n - p)$ is 1.02 when $p = 1$, 1.041 when $p = 2$, 1.062 when $p = 3$, 1.222 when $p = 10$, and so on. Thus, when we increase the order of an AR($p$) process, the steady decrease of $\sigma_M^2$ will be compensated to an increasingly greater extent by the penalty factor. As a result, the decrease of the FPE will be reversed at some point, and this is the point where the FPE is smallest and where we theoretically determine the order of the AR process. In practice, instead of searching for the order of an AR($p$) process for which the FPE is smallest, we select several competing models, such as AR(1) and AR(2), by visually inspecting ACF and PACF correlograms. Then we compare the FPE values of these selected models and select the one whose FPE is smallest.

## [4] AIC, AICc, BIC, and SBC

Although the FPE is easily calculated, it can be used as a selection criterion only for an AR($p$) process. Several statistics utilize the $-2$ log-likelihood ratio and can be used more generally as selection criteria for ARMA($p, q$) models. Akaike's Information Criterion (AIC) consists of two parts (Akaike 1973):

$$AIC(p, q) = -2LL + 2(p + q + 1)$$

where LL is the log-likelihood of an ARMA($p, q$)process.

The first part is the –2 log-likelihood value. The second part, $2(p + q + 1)$, is a penalty factor for a high model order. As we increase the order of our model, the –2LL value will decrease, but its decrease will be compensated by the increasing value of the penalty factor. We will select the model that has the smallest AIC.

The $AIC(p, q)$ has an over-fitting tendency, especially when $p$ and/or $q$ are large relative to $n$. The $AICc(p, q)$ is a corrected version of the $AIC(p, q)$ (Burnham and Anderson 2004; Hurvich and Tsai 1989; Shibata 1976). The $AICc(p, q)$ contains a stronger penalty factor:

$$AICc(p,q) = -2LL + 2(p+q+1)\frac{n}{n-p-q-2}$$

where $n$ is the number of cases.

As $n$ increases, $\dfrac{n}{n-p-q-2}$ gets closer to 1, and the difference between $AIC(p, q)$ and $AICc(p, q)$ will decrease.

Schwarz's Bayesian Criterion, $SBC(p, q)$, and the Bayesian Information Criterion, $BIC(p, q)$, are other variants of $AIC(p, q)$ to correct its over-fitting tendency (Schwarz 1978; Akaike 1979):

$$SBC(p,q) = -2LL + (p+q+1)\ln(n)$$

$$BIC(p,q) = -2LL - (n-p-q)\ln(1-\frac{p+q}{n}) + (p+q)\ln(n) + (p+q)\ln\{\frac{\sigma_Y^2 - \sigma_M^2}{\sigma_M^2(p+q)}\}$$

where $\sigma_M^2$ is the residual variance of an $ARMA(p, q)$ process and $\sigma_Y^2$ is the variance of the residuals, $\{Y_t\}$.

Statistical programs report different selection criteria.[19] Using a calculator or statistical program's compute procedure, we can calculate all these criteria. Most statistical programs report the residual variance and the log-likelihood value. AIC, AICc, BIC, and SBC are based on the log-likelihood value. FPE utilizes the residual variance. For $\sigma_Y^2$ in BIC, we can let statistical programs report descriptive statistics for the residuals, $\{Y_t\}$.

When referring to these model selection criteria, we need to consider the following points. First, these selection criteria are based on the residual variance, and thus they are appropriate when the residual distribution is normal.

The normality of residual distribution can be checked with the histogram of residuals and the normal Q–Q or normal P–P plots (see chapter 3). Second, model selection criteria are only of complementary use. We select an ARMA($p$, $q$) model according to the visual inspection of ACFs and PACFs. Then we can compare the selected model and other competing models, checking whether there is an obviously better one in terms of the selection criteria.

## B. ARCH(q) and GARCH(q, p) Models

In time series analysis, we assume constant variance (homoscedasticity) of residuals. However, the variance of residuals may not be constant over time. Non-constant variance of residuals can take various forms. It may consistently increase or decrease with an increasing level of process (*heteroscedasticity*). We usually regard heteroscedastic variance as a simple characteristic of our data with no theoretical implications. Heteroscedasticity may bias the size of the standard errors of estimated coefficients. Therefore, we control serious heteroscedasticity by transforming original time series (see section 2C).

Non-constant variance may not consistently increase or decrease. It may fluctuate instead, being characterized by the eruptions of high variance in some periods. This type of non-constant variance is termed *conditional heteroscedasticity*. Conditional heteroscedasticity does not necessarily imply changes in the expected mean of a time series and in its long-run variance. With the presence of conditional heteroscedasticity, the long-run variance may be constant.

Conditional heteroscedasticity may indicate an underlying process that is theoretically interesting in the social sciences. For example, macro-level party identification and presidential approval can be more volatile and thus less predictable during certain periods. In addition, the relative effect of an independent variable on macro-level party identification and presidential approval may change across time.

We can take two approaches to deal with conditional heteroscedasticity. One is to respecify our model because conditional heteroscedasticity may be caused by excluded independent variables. The other is to model conditional heteroscedasticity. Autoregressive conditional heteroscedasticity (ARCH) models and generalized autoregressive conditional heteroscedasticity (GARCH) models were developed to estimate conditional heteroscedasticity (Engle 1982; Bollerslev 1986).

ARCH($q$)/GARCH($q,p$) models are similar to AR($p$)/ARMA($p,q$) models. As explained above, an ARMA($p, q$) model is designed to sort out systematic patterns from residuals so that only white noise with a zero mean and a constant variance, WN($0, \sigma^2$), is left over. ARCH($q$)/GARCH($q,p$) models are designed to sort out systematic patterns from the variance of residuals. The difference between ARCH($q$)/GARCH($q,p$) models and AR($p$)/ARMA($p, q$) models is that we utilize squared residuals to identify the former because the variance is the average of the squared differences from the mean. An ARCH($q$) or GARCH($q, 0$) process represents an AR process of the squared residuals, while an AR($p$) or ARMA($p, 0$) process represents an AR process of the residuals.

Another difference between ARCH($q$)/GARCH($q,p$) models and AR($p$)/ARMA($p, q$) models is that we estimate ARCH($q$)/GARCH($q, p$) models after we whiten residuals, while we estimate AR($p$)/ARMA($p, q$) models before we whiten residuals. By the time we apply ARCH($q$)/GARCH($q, p$) models to a time series, the time series should not contain an ARMA($p, q$) process as well as a trend and seasonality. Therefore, a time series with conditional heteroscedasticity that can be represented by an ARCH($q$) or GARCH($q,p$) model is white noise with a zero mean and a long-run constant variance. However, this time series is not IID noise because its current variance is dependent on previous variance(s).

In an ARCH($q$) process, the letter $q$ denotes the number of the previous squared residuals $(Z_{t-1}^2, \ldots, Z_{t-q}^2)$ that are used to explain the current residual variance. When an ARCH($q$) process exists, the current variance $V_t$ is a function of the previous $q$ squared residuals that behaves like an AR($p$) process:

(6) $\quad V_t = b_0 + b_1(Z_{t-1}^2) + b_2(Z_{t-2}^2) + \ldots + b_q(Z_{t-q}^2) = b_0 + \sum_{i=1}^q b_i Z_{t-i}^2$

where $b_i$ is the coefficient of a lag-$i$ previous squared residual, $Z_{t-i}^2$, and $i = 1, 2, \ldots, q$.

Therefore, when an ARCH($q$) process is present, $Z_{t+1}$ is not always equally predictable with the same mean (white noise). For example, for an ARCH(1) process, $V_t = b_0 + b_1(Z_{t-1}^2)$, the increase of variance caused by the increase of the current squared residuals will persist by the ratio of $b_1$. Therefore, when the current variance is higher, $Z_{t+i}$ ($i \geq 1$) will be less predictable until the increase of the current residual variance virtually disappears.

A GARCH($q, p$) model uses past variances, along with past squared residuals, to model the current residual variance. If the conditional variance is explained by its past variances, the number of $q$ terms required to explain the current variance may reduce and the model can be more parsimonious than the ARCH($q$) model (Bollerslev 1986). The letter $p$ denotes the number of the previous variances ($V_{t-1}, \ldots, V_{t-p}$) that are used to explain the current residual variance. When a GARCH($q, p$) process exists, the residuals are a realization of white noise that is characterized by a conditionally non-constant variance that behaves like an ARMA($p, q$) process:

$$(7) \quad V_t = b_0 + \left\{ b_{1e} Z_{t-1}^2 + b_{2e} Z_{t-2}^2 + \ldots + b_{qe} Z_{t-q}^2 \right\} + \left\{ b_{1b} V_{t-1} + b_{2b} V_{t-2} + \ldots + \right.$$

$$\left. b_{pb} V_{t-p} \right\} = b_0 + \sum_{i=1}^{q} b_{ie} Z_{t-i}^2 + \sum_{j=1}^{p} b_{jb} V_{t-j}$$

where $b_{ie}$ is the coefficient of a lag-$i$ previous squared residual, $Z_{t-i}^2$; $b_{jb}$ is the coefficient of a lag-$j$ previous residual variance, $V_{t-j}$; $i = 1, 2, \ldots, q$; and $j = 1, 2, \ldots, p$.

We should note that the roles of the $p$ and $q$ terms are reversed in ARCH($q$)/ GARCH($q, p$) models, compared with AR($p$)/ARMA($p, q$) models (see section 5A). Like an MA($q$) model in which we explain the current residual ($Y_t$) with current and past white noise components ($Z_t, Z_{t-1}, \ldots, Z_{t-q}$), we explain the current variance ($V_t$) with past squared white noise components. Thus, we use $q$ to denote the order of the ARCH process. Since squared residuals are autocorrelated, however, the $q$ term, $\sum_{i=1}^{q} b_i Z_{t-i}^2$, is similar to the $p$ term of the AR process.

Like an AR($p$) model in which we explain the current residual ($Y_t$) with its past values ($Y_{t-1}, \ldots, Y_{t-p}$), we explain the current variance ($V_t$) with its past values, along with past squared white noise components. Thus, we use $p$ to denote the order of the GARCH process. Since variances of white or IID noise are uncorrelated in the long run, however, the $p$ term, $\sum_{j=1}^{p} b_j V_{t-j}$, indicates how many previous uncorrelated variances are used to explain the current variance. Therefore, it is similar to the $q$ term of the MA process that indicates how many uncorrelated previous white noise components are used to explain the current value.[20]

Since GARCH($q, p$) models are similar to ARMA($p, q$) models, we can interpret a GARCH($q, p$) process and an ARMA($p, q$) process similarly. The coefficients $b_i$ and $b_j$, if they are significant, indicate that the short run variance is conditioned on past shocks that persist (AR process of the squared residuals) and its own uncorrelated previous values (MA process of the conditional variance in the previous $p$ periods), respectively. For example, with an estimated GARCH(1, 1) model, we can discuss whether an event leads to a greater volatility of a time series and whether the volatility, once sparked, tends to persist beyond the event that sparked it (Gronke and Brehm 2002). If the residual variances do not depend on the conditional variance in the previous $p$ periods ($b_j = 0$), a GARCH($q, p$) process is an ARCH($q$) process.

Once we have whitened and squared residuals, the procedure through which we identify ARCH($q$)/GARCH($q, p$) models is similar to the procedure we use to identify an ARMA($p, q$) model. First, we need to inspect the residual plots. We can be more generous about the constant-variance assumption in the social sciences than in the natural sciences or in economics. No one expects the variance to remain unchanged all the time when dealing with human behaviors and social or political phenomena. If residual variances remain fairly stable across time, we may not need to move to the ARCH($q$)/GARCH($q, p$) models. However, if residual variances increase substantially in some time periods and if we believe that such an increase is theoretically meaningful, then we need to model the non-constant variance to complete our model of the time-dependent behavior of a time series.

To model conditional heteroscedasticity, we first square the whitened residuals. For an ARCH($q$) process, we can identify an AR($p$) process in the squared residuals, $\{Z_t^2\}$, by referring to their ACF and PACF correlograms (see section 5A). Decaying autocorrelations of the squared residuals and partial autocorrelations with $q$ spikes will indicate the presence of an ARCH($q$) process in the residual variance. A GARCH($q, p$) process will produce ACF and PACF correlograms similar to those for an ARMA($p, q$) process.

Once we detect the presence of an ARCH($q$) or GARCH($q, p$) process through our visual inspection of ACF and PACF correlograms, we can apply a Lagrange multiplier test to confirm what we find from our visual inspection of squared residuals (Engle 1982). First, we regress squared residuals on the lagged squared residuals. This regression takes the same form as equation (6). Second, we take the coefficient of determination ($R^2$) from the regression. Then, we calculate the Lagrange multiplier as:

$$LM = nR^2$$

where $n$ is the total number of observations.

If there are no ARCH effects, the individual coefficient estimates should not be significantly different from zero and the model should be insignificant.

LM is well approximated by the chi-squared distribution with $q$ degrees of freedom. If LM is larger than the corresponding chi-squared value, we reject the null hypothesis of no ARCH effects.[21] For a GARCH$(q, p)$ model, we can utilize a Lagrange multiplier test for an ARCH$(q + p)$ model (Bollersley 1986). For example, if the LM statistic rejects the null hypothesis of no ARCH(2) effect, we can also reject the null hypothesis of no GARCH(1, 1) effect.

With ARCH$(q)$/GARCH$(q, p)$ models in univariate time series analysis, we can discuss whether residual variance significantly varies across time. In multiple time series analysis, we can explain changes in the relative effect of an independent variable on the dependent variable as well as the conditional non-constant residual variance.

In the social sciences, however, simply knowing that residual variance is volatile may not be enough. An ARCH$(q)$ process does not convey a meaningful theoretical implication, just like an AR$(p)$ process. In the social sciences, we are usually expected to explain what factors cause the volatility of residual variance, rather than simply to report the volatility and its systematic pattern. For example, variance in presidential approval or macro-level party identification may decrease in election years but increase as party identification becomes weaker (Gronke and Brehm 2002; Maestas and Preuhs 2000). To explore this relationship, we transform an ARCH$(q)$ or GARCH$(q, p)$ model into a multiple time series analysis form.[22] The dependent variable will be the squared residuals that represent the residual variance. The right-hand side of the equation will include explanatory variables that are selected according to our theory, in addition to $q$ terms and/or $p$ terms of a GARCH$(q, p)$ model. For example, Gronke and Brehm (2002) extend an ARCH(1) model to include various dummy variables that represent various events (such as re-election year, election year, adverse economic event, administration scandal, etc.), transforming the original ARCH(1) model into a multiple time series analysis model. With this model, they control the AR process in the dependent variable's residual variance and discuss whether each of these events causes the volatility of the residual variance in presidential approval.

### C. The Case of Monthly Violent Crime Rates, 1983–1992

To model an ARMA$(p, q)$ process in the deseasonalized and detrended crime rates, we first visually inspect the sample ACF of residuals. In figures 21 and 22, two points draw attention. One is that sample ACFs show a rapid decay. The other is that PACFs become negligible after lag 2. These two points indicate that

TABLE 4. Preliminary Parameter Estimates of AR(2) Model and Two Other Competing Models for the Deseasonalized and Detrended Monthly Violent Crime Rates, 1983–1992

| | Coefficient | | Selection criteria | | |
|---|---|---|---|---|---|
| | | | $AICc$ | $\sigma_M^2$ | $-2LL$ |
| AR(2) | $\phi_1$ | .53 | 423.68 | 1.89 | 417.47 |
| | $\phi_2$ | .22 | | | |
| AR(1) | $\phi$ | .67 | 430.18 | 2.03 | 426.08 |
| ARMA(1,1) | $\phi$ | .85 | 426.73 | 1.94 | 420.52 |
| | $\theta$ | -.05 | | | |

NOTE. Coefficients were estimated from the monthly violent crime rates that were deseasonalized by the additive method and then detrended. $\sigma_M^2$ is the residual variance. $-2LL$ is $-2$ log-likelihood.

an AR process of order 2 is present in the residuals. In figure 22, however, the PACF at lag 2 does not lie significantly far outside the confidence bounds. Thus, an AR(1) process may also nicely describe the stationary process of the residuals.

Once we select an AR(2) process, statistical programs preliminarily estimate the coefficients of the AR(2) process, $\phi_1$ and $\phi_2$, using a particular algorithm. Then they use these estimates as starting points of the optimization process. Table 4 presents the preliminary estimates of the AR(2) model, along with those of two other competing models.[23]

Table 4 reports only three selection criteria because we do not need to refer to many selection criteria at the preliminary estimation stage. As we have already determined through visual inspection of ACF and PACF correlograms (figures 21 and 22), the AR(2) model appears to be better than the other two models in representing the stationary process of the deseasonalized and detrended monthly violent crime rates. According to table 4, among the three models, the AICc is smallest for the AR(2) model. The AR(2) model also has the smallest residual variance and $-2$ log-likelihood value. However, the other two models are not significantly worse than the AR(2) model.

Once we obtain a preliminary estimate of a selected ARMA($p, q$) model, the next step is to optimize the preliminary estimate with the maximum likelihood estimation method. In most multipurpose programs, the preliminary estimation procedure and the maximum likelihood estimation procedure are integrated. Therefore, we actually begin the optimization process by inputting values for $p$ and/or $q$ of the ARMA($p, q$) process, without caring about the preliminary estimates.[24] In the case of the deseasonalized and detrended monthly violent crime rates, we input 2 for $p$ and 0 for $q$.

TABLE 5. Maximum Likelihood Estimates for the Deseasonalized and Detrended Monthly Violent Crime Rates, 1983–1992

| | Coefficient | Standard error | Selection criteria | | | | | | |
|---|---|---|---|---|---|---|---|---|---|
| | | | $AICc$ | $AIC$ | $BIC$ | $SBC$ | $FPE$ | $\sigma_M^2$ | $-2LL$ |
| AR(2) | $\phi_1$ .54 | .09 | 422.32 | 422.18 | 425.56 | 430.54 | 1.92 | 1.86 | 416.11 |
| | $\phi_2$ .28 | .09 | | | | | | | |
| AR(1) | $\phi$ .73 | .06 | 429.44 | 429.44 | 430.34 | 435.02 | 2.05 | 2.01 | 425.33 |
| ARMA(1,1) | $\phi$ .86 | .06 | 425.18 | 425.05 | 428.42 | 433.42 | n/a | 1.91 | 418.98 |
| | $\theta$ .26 | .12 | | | | | | | |

NOTE. The monthly violent crime rates were deseasonalized by the additive method and then detrended. $\sigma_M^2$ is the residual variance. $-2LL$ is $-2$ log-likelihood. Since FPE is only for a pure AR($p$) process, it is not reported for the ARMA(1, 1) process.

Table 5 presents the maximum likelihood estimates for the three ARMA($p$, $q$) models and the selection criteria that were explained above. Although programs may begin the optimization process with slightly different preliminary estimates, they will produce almost identical maximum likelihood estimates.[25]

According to table 5, all coefficients are statistically significant, indicating that they are contributing to the overall fit of each model. We already found through the visual inspection of ACF and PACF correlograms that either an AR(2) process or an AR(1) process represents the systematic pattern of the deseasonalized and detrended monthly violent crime rates. All selection criteria indicate that the AR(2) model is better than the AR(1) and ARMA(1, 1) models in representing the stationary process of the deseasonalized and detrended crime rates. In addition, the difference of $-2LL$ between the AR(1) and AR(2) models exceeds the chi-squared value (3.8415) with one degree of freedom at $\alpha = .05$. Therefore, we determine that the AR(2) model represents the systematic pattern of residuals better than does the AR(1) model.

In sum, according to the visual inspection of residuals (figures 21 and 22) and the selection criteria (table 5), we select the AR(2) model and take the maximum likelihood estimates of its two coefficients:

$$Y_t = 0.54Y_{t-1} + 0.28Y_{t-2} + Z_t$$

where $Y_t$ is the deseasonalized and detrended crime rates, $Y_{t-1}$ and $Y_{t-2}$ are the lag-1 and lag-2 previous values of $Y_t$, respectively, and $Z_t$ is white noise with a mean equal to 0 and a variance equal to 1.9.

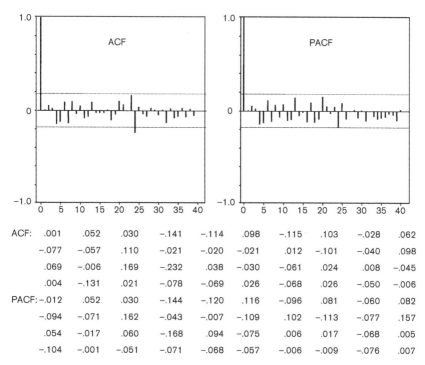

| ACF: | .001 | .052 | .030 | −.141 | −.114 | .098 | −.115 | .103 | −.028 | .062 |
|---|---|---|---|---|---|---|---|---|---|---|
| | −.077 | −.057 | .110 | −.021 | −.020 | −.021 | .012 | −.101 | −.040 | .098 |
| | .069 | −.006 | .169 | −.232 | .038 | −.030 | −.061 | .024 | .008 | −.045 |
| | .004 | −.131 | .021 | −.078 | −.069 | .026 | −.068 | .026 | −.050 | −.006 |
| PACF: | −.012 | .052 | .030 | −.144 | −.120 | .116 | −.096 | .081 | −.060 | .082 |
| | −.094 | −.071 | .162 | −.043 | −.007 | −.109 | .102 | −.113 | −.077 | .157 |
| | .054 | −.017 | .060 | −.168 | .094 | −.075 | .006 | .017 | −.068 | .005 |
| | −.104 | −.001 | −.051 | −.071 | −.068 | −.057 | −.006 | −.009 | −.076 | .007 |

FIGURE 26. Sample ACF and PACF of the residuals, $\{z_t\}$, obtained by fitting AR(2) process to $\{y_t\}$.

After fitting an AR(2) model to the deseasonalized and detrended crime rates, $\{Y_t\}$, we can check the appropriateness of the model by visually inspecting the ACF and PACF correlograms of the rescaled residuals, $\{Z_t\}$. According to figure 26, the ACF and PACF correlograms indicate that the rescaled residuals are a realization of white noise. No apparent systematic patterns are observed in the figure. This point can be double-checked by inspecting the scatter plots of the residuals, $\{Z_t\}$. The plots show much less smoothness, indicating that the dependency on time has almost disappeared (figure 27). There seems to be no positive or negative correlation among residuals. Also, no obvious deviations from the assumptions of zero mean and constant variance are observed.

As a complementary check, we can go back to the randomness tests and see whether what are left over in the series, $\{Z_t\}$, are random components. By visually inspecting ACFs and PACFs of the deseasonalized and detrended crime rates (figures 21 and 22) and by referring to the selection criteria of the

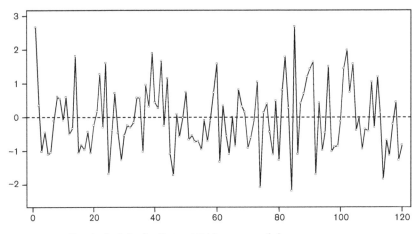

FIGURE 27. Residuals, $\{z_t\}$, after fitting AR(2) process to $\{y_t\}$.

maximum likelihood estimates (table 5), we have already found that the AR(2) model is better than the AR(1) and ARMA(1, 1) models in representing the systematic pattern of the deseasonalized and detrended crime rates. Thus, at this stage, it will be enough to conduct the randomness tests only for the residuals from which the AR(2) process has been taken out. To compare the performance of the AR(2) model with that of the other two models, table 6 presents the randomness test results for the residuals of all three models. All three models passed most tests. The AR(1) model did the worst, but it still passed four out of six tests. The AR(2) and ARMA(1, 1) models passed five out of six tests. Visual inspection of the ACFs and PACFs of the deseasonalized and detrended crime rates, however, clearly indicates the presence of the AR(2) process in residuals. Since the residual distribution from which the AR(2) process has been taken out passed five out of six randomness tests at $\alpha = .05$ and passed all six tests at $\alpha = .01$, we can confidently conclude that only random components are left over. In addition, the Durbin-Watson $d$ statistic, 1.88, is greater than the upper bound of the Durbin-Watson $d$ statistic, 1.502, when the level of significance is .01, and 1.715, when the level of significance is .05. Therefore, we do not reject the null hypothesis and conclude that a positive serial correlation is no longer present in residuals.

We estimated the trend and seasonality of the monthly violent crime rates from 1983 to 1992. Also, we identified the systematic pattern of the desea-

TABLE 6. Randomness Test Statistics for the Residuals from Which the Estimated Systematic Pattern Has Been Removed

| Test | Observed value | | | Critical value | IID hypothesis | | |
|------|------|------|------|------|------|------|------|
| | AR(2) | AR(1) | ARMA(1,1) | at α = .05 | AR(2) | AR(1) | ARMA(1,1) |
| Ljung-Box | 30.58 | 52.99 | 36.58 | 31.41 | NR | R | R |
| Mcleod-Li | 11.93 | 15.95 | 11.26 | 33.92 | NR | NR | NR |
| Turning points | 0.51 | 1.82 | 1.82 | 1.96 | NR | NR | NR |
| Difference sign | 2.36 | 2.05 | 1.73 | 1.96 | R* | R | NR |
| Rank | 0.18 | 0.15 | 0.18 | 1.96 | NR | NR | NR |
| Runs | −0.41 | 1.13 | −0.41 | ±1.96 | NR | NR | NR |

NOTE. The Ljung-Box and Mcleod-Li tests were calculated with the residuals from lag 1 to lag 24. The other tests were calculated with all the residuals. R: rejected; NR: not rejected.
* The IID hypothesis is not rejected at the .01 level (the critical value is 2.58).

sonalized and detrended series. The complete time series model of the monthly violent crime rates, $\{X_t\}$, from January 1983 to December 1992 is

$$X_t = 41.96 + S_t + 0.19\text{Time} + 0.54Y_{t-1} + 0.28Y_{t-2} + Z_t$$

where Time = 1, 2, ..., 120;
$S_t$ = −2.72 (January), −6.93 (February), −2.16 (March), −2.70 (April), 0.90 (May), 1.79 (June), 5.39 (July), 5.86 (August), 2.15 (September), 1.96 (October), −2.51 (November), and −1.05 (December); and
$Z_t$ = WN(0, 1.9).

According to this final model, the monthly violent crime rate tended to increase by 0.19% on average every month from 1983 to 1992. In each year, the violent crime rate tended to show monthly variations, with the lowest rate in February (lower than the year's violent crime rate by 6.93% on average) and the highest rate in August (higher than the year's violent crime rate by 5.86% on average).

After these trend and seasonal components were removed from the original crime rates, the residuals still contained a systematic pattern. The current deseasonalized and detrended crime rates tended to be positively related to the lag-1 and lag-2 previous observations. Once this AR(2) process is removed from the deseasonalized and detrended monthly crime rates, only white noise is left.

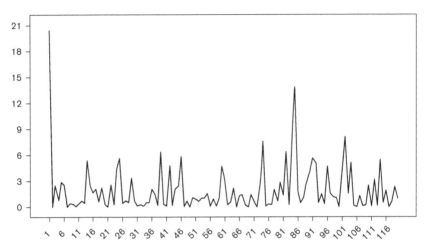

FIGURE 28. Squared residuals, $\{z_t^2\}$.

We may finish our modeling of the time-dependent behavior at this point, as the monthly violent crime rates do not contain discernible systematic patterns. According to figure 27, the dependency on time has now disappeared. According to figure 27, however, the variances in the period between the 81st time point (October 1989) and the 84th time point (January 1990) seem to be larger, compared with the variances in the other period. On the other hand, the variances in the periods between the 7th time point (August 1983) and the 12th time point (January 1984), between the 14th time point (March 1984) and the 23rd time point (December 1984), and between the 50th time point (March 1987) and the 59th time point (December 1987) seem to be smaller, compared with the variances in the other period. Therefore, the variance of $\{Z_t\}$ may be volatile, although it appears to be constant in the long run.

To check the variance of $\{Z_t\}$ more clearly, we squared $\{Z_t\}$. Figure 28 presents the plots of the squared residuals. It clearly shows the possible volatility of the variance of $\{Z_t\}$. However, according to the sample ACF and PACF correlograms of the squared residuals, no discernible GARCH$(q, p)$ process is identified (figure 29). The LM test confirms that we do not need to model conditional heteroscedasticity. The LM statistic for an ARCH(1) model is 0.372 ($0.0031 \times 120$), which is smaller than the chi-squared value at $\alpha = .05$ with one degree of freedom (3.84). The LM statistic for an ARCH(2) model is 1.932 ($0.0161 \times 120$), which is also smaller than the chi-squared value at $\alpha = .05$ with two degrees of freedom (5.99).[26] Therefore, an

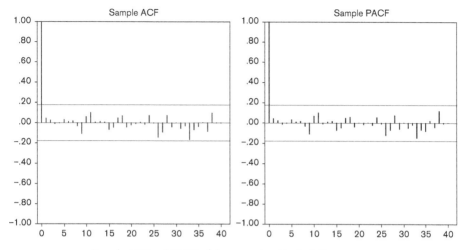

FIGURE 29. Sample ACF and PACF of the squared residuals, $\{z_t^2\}$.

ARCH(1) process, an ARCH(2) process, or a GARCH(1, 1) process does not exist in the whitened monthly violent crime rates.

## (6) FURTHER READING

Akaike, H. 1969. "Fitting Autoregressions for Predictions." *Annals of the Institute of Statistical Mathematics* 21:243–247.

———. 1973. "Information Theory and an Extension of the Maximum Likelihood Principle." In *The Second International Symposium on Information Theory*, ed. B. N. Petrov and F. Csaki, 267–281. Budapest: Akademial kiado.

———. 1979. "A Bayesian Extension of the Minimum AIC Procedure of Autoregressive Model Fitting." *Biometrika* 66:237–242.

Bauwens, L., S. Laurent, and J. V. K. Rombouts. 2006. "Multivariate GARCH Models: A Survey." *Journal of Applied Econometrics* 21:79–109.

Bollerslev, T. 1986. "Generalized Autoregressive Conditional Heteroskedasticity." *Journal of Econometrics* 31:307–327.

Box, G. E. P., and D. R. Cox. 1964. "An Analysis of Transformations." *Journal of the Royal Statistical Society* 26:211–243.

Box, G. E. P., and D. Pierce. 1970. "Distribution of Autocorrelations in Autoregressive Moving Average Time Series Models." *Journal of the American Statistical Association* 65:1509–1526.

Burnham, K. P., and D. R. Anderson. 2004. "Multimodel Inference: Understanding AIC and BIC in Model Selection." *Sociological Methods and Research* 33:261–304.

Durbin, J., and G. S. Watson. 1951. "Testing for Serial Correlation in Least Squares Regression II." *Biometrika* 38:159–178.

Engle, R. F. 1982. "Autoregressive Conditional Heteroskedasticity with Estimates of the Variance of United Kingdom Inflation." *Econometrica* 50:987–1007.

Goldfield, A., and R. Quandt. 1965. "Some Tests for Heteroscedasticity." *Journal of American Statistical Association* 60:539–547.

Green, D. P., A. S. Gerber, and S. L. De Boef. 1999. "Tracking Opinion over Time: A Method for Reducing Sampling Error." *Public Opinion Quarterly* 63:178–192.

Gronke, P., and J. Brehm. 2002. "History, Heterogeneity, and Presidential Approval: A Modified ARCH Approach." *Electoral Studies* 21:425–452.

Kohn, R., and C. Ansley. 1986. "Estimation, Prediction, and Interpolation for ARIMA Models with Missing Data." *Journal of the American Statistical Association* 81:751–761.

Ljung, G. M., and G. E. P. Box. 1978. "On a Measure of Lack of Fit in Time Series Models." *Biometrika* 65:297–303.

Maestas, C., and R. Preuhs. 2000. "Modeling Volatility in Political Time Series." *Electoral Studies* 19:95–110.

McLeod, A. I., and W. K. Li. 1983. "Diagnostic Checking ARMA Time Series Models Using Squared Residual Autocorrelations." *Journal of Time Series Analysis* 4:269–273.

Schwarz, G. 1978. "Estimating the Dimension of a Model." *Annals of Statistics* 6:461–464.

Shibata, R. 1976. "Selection of the Order of an Autoregressive Model by Akaike's Information Criterion." *Biometrika* 63:117–126.

# THREE

# *Diagnostics*

## (1) RESIDUAL ASSUMPTIONS

In regression analysis, residuals are what were left after we removed a slope from cross-sectional data. Residuals in time series analysis are what were left after we removed seasonal components and a trend from time series data. If the residuals contain a systematic pattern, the pattern should also be taken out.

In regression analysis, several properties of residuals should be satisfied for a slope estimate to be BLUE (best linear unbiased estimate):

(a)  a zero mean

(b)  a constant variance, $\sigma^2$, for all the values of the independent variable

(c)  no autocorrelation

(d)  a normal distribution.

Univariate time series analysis is bivariate regression analysis with Time as the independent variable. Therefore, residuals should also satisfy these properties, if a fitted model is appropriate. In univariate time series analysis, we do not need to check the first two properties. When we estimate and eliminate seasonality, trends, and systematic patterns of residuals, we take care of violations of a zero mean and a constant variance (see chapter 2). The third and fourth properties deserve our attention.

### A. No-Autocorrelation Assumption

Different from cross-sectional data, data collected across time frequently contain serially correlated noise. In this case, an important assumption of

OLS regression analysis, no autocorrelation in residuals, is violated. In time series analysis, we will know whether the no-autocorrelation assumption is violated by the end of the time series modeling process, as shown in chapter 2.

With the presence of autocorrelation in the residuals, positive residuals and negative residuals cancel out each other (Berry 1993). Therefore, our trend estimate will be unbiased in the long run and can be used to explain and forecast the temporal behavior of a time series. However, positive (negative) autocorrelation among residuals reduces (increases) the standard error of the trend estimate, and thereby the statistical significance of the trend estimate may be overestimated (underestimated). Positive (negative) autocorrelation among residuals may also reduce (increase) the mean squared error of the regression equation and consequently inflate (deflate) the $F$ statistic. Therefore, the overall level of significance of the regression model may be overestimated (underestimated).

This problem may be limited, if the trend estimate is highly significant or insignificant. If highly significant, even without the deflated standard error, the trend estimate is likely to remain significant. If highly insignificant, even without the inflated standard error, the trend estimate is likely to remain insignificant. However, when the trend estimate is weakly significant or insignificant, with a level of significance near .05, we need to be concerned about the possibility that the autocorrelation in residuals may make the actually insignificant (significant) trend estimate significant (insignificant).

A commonly employed remedy for cross-sectional data with autocorrelated residuals is generalized least squares regression analysis. When employing this method for a time series analysis, however, the whole series including residuals is transformed. As a result, the estimated trend and systematic pattern of residuals are for the transformed series, not for the original one. In addition, the forecasted values will be the forecasted values of the transformed series.

In univariate time series analysis, we do not need to transform data when we find autocorrelation in residuals. When we identify a significant autocorrelation in residuals, we may need to return to the trend estimation stage after we estimate and eliminate the systematic pattern of residuals. This process is called *prewhitening* because we make the variable's disturbances consist of white noise (see chapter 6, section 2). We can also remove the systematic pattern of residuals by smoothing a time series (see chapter 5). We re-estimate a trend in the prewhitened or smoothed time series and check whether the trend estimate remains identical and statistically significant (or insignificant). Because the autocorrelation of residuals does not bias trend estimates, a newly estimated trend coefficient is expected to be similar to the old one, but it may be no longer statistically significant or insignificant. In the case of the monthly violent crime rates, the newly estimated trend is 0.188 ($p = .004$).

Autocorrelation can be caused by an incorrect functional form of an independent variable. With regard to the monthly violent crime rates, this is certainly not the case because visual inspection supports the linearity of trend. If we find a serious autocorrelation in residuals, however, we need to check whether we appropriately identified the functional form of trend.

In bivariate or multiple time series analysis, we can adjust estimated coefficients and their standard errors, following the procedure proposed by Cochrane and Orcutt (1949) or Prais and Winsten (1954)—see chapter 6. In univariate time series analysis, this procedure is not appropriate in that Time is not an observed variable. It simply indicates the observation order of the values of the dependent variable. The estimated trend in univariate time series analysis depicts the average change of the dependent variable associated with the level of a regular time interval, not the impact of the independent variable, Time, on the dependent variable.

The no-autocorrelation assumption is related to the trend estimation, not to seasonality. To estimate seasonality accurately, we should estimate it before we eliminate systematic patterns from residuals or smooth a time series. Seasonality consists of cyclical regular fluctuations, and deseasonalizing a time series helps us estimate trends and systematic patterns of residuals more accurately. Eliminating systematic patterns from residuals will damp the original time series, the same consequence of smoothing a time series (see chapter 5). Therefore, seasonal fluctuations will change after we eliminate systematic patterns from residuals. This is why we estimate and eliminate seasonal components before we estimate trends and systematic patterns of residuals (see chapter 2).

## B. Normality Assumption

In OLS regression analysis, the assumption that residuals are normally distributed is mainly related to the interpretation of slope coefficients and their significance tests (Berry 1993; Fox 1991). In time series analysis, the normality assumption is related to the interpretation of trends and their significance tests. With the non-normality of residual distribution, we may not appropriately test whether the estimated coefficient is significantly different from zero. In addition, violation of the normality assumption can reduce the robustness of a trend estimate. As in OLS regression analysis, the non-normality of residual distribution may indicate that some excluded independent variables exert significant influence over a trend estimate, and thus the estimated coefficient may not be BLUE. It may also indicate that an estimated functional form of trend is inaccurate. For instance, we may need to estimate a nonlinear trend.

The normality assumption tends to draw relatively greater attention in time series analysis than in OLS regression analysis because the number of cases is usually small in time series analysis. We may not need to care much about the normality assumption if our estimation is based on a large sample (Lewis-Beck 1980; Berry 1993).

Non-normality tends to go hand in hand with non-constant variance and/or autocorrelation. Thus, if we transform a time series to stabilize its variance and/or if we remove autocorrelation from residuals, residual distribution is often normalized as well. This is one reason that we check the normality of distribution at the end of a time series modeling procedure.

The normality of distribution can be checked directly with *skewness* and *kurtosis* statistics. If a distribution is normal, it should not be highly skewed either in the right direction (longer right tail, positively skewed) or in the left direction (longer left tail, negatively skewed). In addition, it should not be highly kurtic, either in the positive direction (more peaked than the normal distribution, leptokurtic) or in the negative direction (less peaked than the normal distribution, platykurtic). Skewness and kurtosis statistics are provided by most statistical programs. For a reasonably normal distribution, both statistics should near zero. Conventionally, if their absolute values are less than or equal to 3, we do not carry out transformation to correct the non-normality.

We can employ the Anderson-Darling test, the one-sample Kolmogorov-Smirnov test, the Ryan-Joiner test, and the Shapiro-Wilk test to check the normality of distribution (Razali and Wah 2011; Seier 2002). These tests compare the observed residual distribution with the normal distribution to check whether the observed cases came from the normal distribution with a mean of zero and a variance of one. If these test statistics are statistically insignificant, we cannot reject the null hypothesis that residuals are normally distributed.

These test statistics, however, do not convey enough information on residual distribution. As Fox (1991) recommends, visual inspection of residual plots can be more helpful when we check the normality of residual distribution in most social science studies. In practice, there will be a few social science datasets that show perfect or near-perfect normal distribution. Residuals may slightly fail the normality tests. However, their distribution may look reasonably normal. For instance, residual distribution may be unimodal and not highly skewed. In this case, we may not care much about the failed tests. This is the case especially when the sample size is large.

We can visually check the normality of residual distribution with a scatter plot, a stem-and-leaf plot, a histogram, a normal Q–Q (quantile–quantile) plot, or a normal P–P (probability–probability) plot. For visual inspection of residuals, it is helpful to standardize them. To standardize residuals, we divide them by an estimated residual standard deviation.

If residuals are normally distributed, a standardized residual scatter plot should resemble that of a normal distribution with a mean of zero and a variance of one. Also, most (conventionally, 95%) of the standardized residuals should fall inside the ±1.96 confidence limits.

A histogram and a stem-and-leaf plot show us the shape of residual distribution, and we can compare it with a normal curve. A histogram is a kind of bar graph, and it conveys rough distributional information. When data are sparse, a histogram can produce misleading impressions of the residual distribution. Fox (1991) recommends smoothing a histogram if our sample size is small, that is if the number of cases is less than 1,000. In time series analysis, smoothing a histogram is helpful because the number of cases is usually several hundred at most. We can smooth a histogram by adjusting the interval of the $x$-axis to contain 5% or more of all the cases on average. For a small sample, that is with less than 100 cases, a stem-and-leaf plot is preferred to a histogram.

In histograms and stem-and-leaf plots, several points deserve our attention because they can indicate serious deviations from the normality assumption (Fox 1991):

(a) a multimodal distribution
(b) the skewness of distribution
(c) obvious outliers.

If the normality assumption is satisfied, a single mode should be observed. An obvious bimodal or multimodal distribution may indicate that we need to qualitatively divide a time series into two or more time periods and analyze them separately.

If a single-peaked distribution is observed, we need to check that the distribution is not highly skewed in either right or left direction. A perfect normal distribution is unexpected in the social sciences. When residual distribution is highly skewed, however, our estimation of a linear trend may not be the best unbiased one. To estimate a linear trend, we employ the least squares estimate method, which is based on the mean, and highly skewed residual distribution indicates that the mean may not be an efficient measure of central tendency. In addition, when residual distribution is highly skewed, we may not appropriately conduct the significance test of an estimated trend coefficient because the test is based on the mean.

We have already checked outliers when visually inspecting the original time series (see chapter 2). The plot of the original time series will convey clear information on the impact of outliers on trend estimation. Visually inspecting residual distribution, we can double-check the existence of obvious outliers. In the histogram of residuals, the presence of outliers will be informed by bars that are separated far away from the others and also by a heavy tail of the normality curve. We can examine the residual plots against the independent variable, Time, or against the dependent variable, that is the original time series.

Outliers in the scatter plot of the original time series are the observations that may distort the trend estimation. Outliers in the scatter plot of residuals are the observations that are poorly represented by an estimated trend. As mentioned in chapter 2, even when we find seemingly obvious outliers, it may not be desirable to remove them from a time series. Instead, as Lewis-Beck (1980) recommends for OLS regression analysis, we need to conduct two separate analyses with such outliers and with imputed or interpolated values that replace outliers and report both results.

To compare residual distribution graphically with the normal distribution, we can employ a normal Q–Q plot or a normal P–P plot. A normal Q–Q plot plots quantiles of residuals against quantiles of the normal distribution. A normal P–P plot plots cumulative proportions of residuals against cumulative proportions of the normal distribution. In a normal Q–Q plot or a normal P–P plot, residuals will cluster around a straight line that represents a perfect correlation between the residual distribution and the normal distribution, if the residual distribution matches the normal distribution well. These two plots, however, do not convey some vital information that can be provided by a histogram and a stem-and-leaf plot. For example, with a normal Q–Q plot or a normal P–P plot, we cannot detect whether the residual distribution is unimodal or multimodal.

## (2) THE CASE OF MONTHLY VIOLENT
## CRIME RATES, 1983–1992

We have estimated and eliminated the seasonal components and trend of the monthly violent crime rates from 1983 to 1992. Also, we have estimated and taken out the systematic pattern from the deseasonalized and detrended crime rates. Now we know that only white noise is left. The whiteness of residuals, $\{Z_t\}$, indicates that the remaining series are not dependent on time but random. Now we are ready to forecast the future crime rates with the estimated model.

Before we complete our time series modeling procedure, we need to conduct some diagnostics to check whether we have appropriately modeled the monthly violent crime rates. We have already checked for a zero mean, a constant variance, and no serial dependency among residuals (see chapter 2). For more confident interpretation and forecast based on the estimated model, we need to check one more point, the normality of residual distribution.

In the case of $\{Z_t\}$, the skewness and kurtosis of the standardized residuals are 0.754 (standard deviation 0.221) and 1.277 (0.438), respectively. Thus, the residual distribution seems to be close to the normal distribution. In addition, one sample Kolmogorov-Smirnov test statistic is 0.07, and its level of significance is .2. Thus, we do not reject the null hypothesis that the residual distribution is normal.

Visual inspection of several graphs also supports the normality of the residual distribution. Figure 30 presents a histogram of the standardized residuals. The histogram was smoothed by setting the number of intervals to 20 so that each interval can contain five percentages (six observations) of the residuals on average. As explained above, this smoothing allows the histogram to avoid conveying misleading impressions when the number of cases is small. If we increase the sharpness of the histogram, for example by increasing the number of intervals to 40, several more cases make a bar significantly higher than the others (figure 31).

We do not expect histograms to be perfect in shape in the social sciences. That is, a bell-shaped curve is not likely to be put exactly on the upper edges of the bars in a histogram (figures 30 and 31). According to figure 30, the normality assumption seems to be reasonably satisfied. There is a single mode. The shape of distribution is close to the bell shape, though not perfect. As the skewness and kurtosis statistics indicate, the residual distribution is

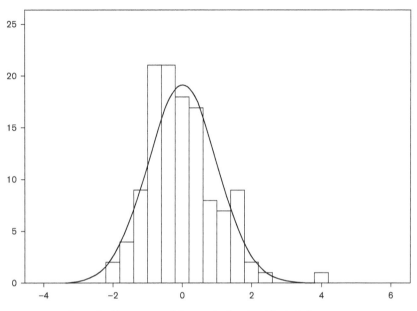

FIGURE 30. Smoothed histogram of the standardized residuals, $\{z_t\}$.

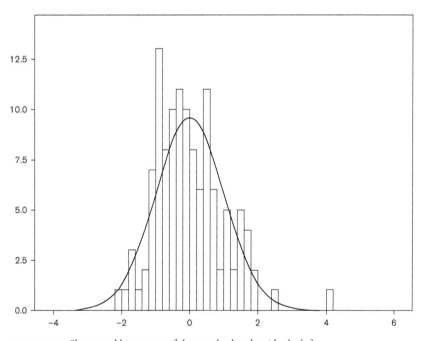

FIGURE 31. Sharpened histogram of the standardized residuals, $\{z_t\}$.

| Stem | Leaf | Frequency |
|------|------|-----------|
| -2. | 0 | 1 |
| -1. | 56679 | 5 |
| -1. | 000001122 | 9 |
| -0. | 555666677778888899999999 | 24 |
| -0. | 0000001111222222333334444444 | 28 |
| 0. | 0001111122333344444 | 19 |
| 0. | 55555567777789 | 14 |
| 1. | 001112344 | 9 |
| 1. | 555667789 | 9 |
| 2. | | 0 |
| 2. | 5 | 1 |

| | | |
|------|------|------|
| Extremes (≥ 4.0) | | 1 |

Stem width: 1.00
Each leaf: 1 case

FIGURE 32. Stem-and-leaf plot of standardized residuals, $\{z_i\}$.

somewhat positively skewed and more peaked than the normal distribution. The residual distribution, however, does not appear to be seriously skewed or leptokurtic. In addition, the distribution seems to be reasonably symmetric: the numbers of positive residuals and negative residuals are similar. There is one outlier on the right tail. This is the crime rate in January, 1983. According to the deseasonalized time series plots (figure 14), however, this outlier was not likely to distort the trend estimation seriously. Thus, it seems to be safe to include this observation in our analysis.

Figure 32 is the stem-and-leaf plot of the standardized residuals. As explained above, when the number of cases is not large enough, that is smaller than several hundreds, stem-and-leaf plots tend to describe the distribution more accurately than histograms. In fact, a stem-and-leaf plot is a sort of smoothed histogram. For instance, the number of intervals is 11 in Figure 32, which is about half the number in figure 30. We can see that the normality of the residual distribution stands out a little more clearly in figure 32 than in figure 30. Figures 30 and 32 convey almost identical information, except that we can directly read the values of cases and the number of cases in each bar in figure 32. In addition, the outlier is not marked as a stem and its leaf, but directly mentioned as an extreme in figure 32.

In figures 33 and 34, the standardized residuals are ordered and matched against corresponding quantiles and cumulative probabilities, respectively, of

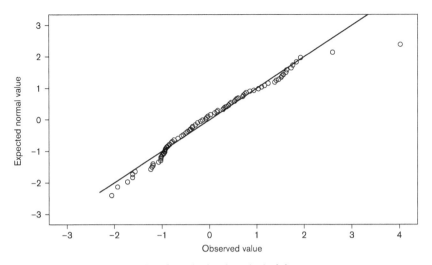

FIGURE 33. Normal Q–Q plot of standardized residuals, $\{z_t\}$.

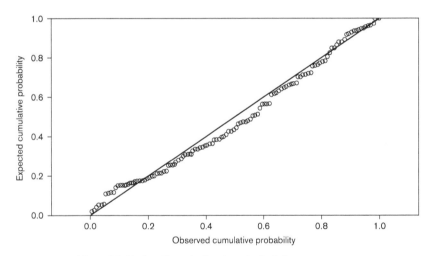

FIGURE 34. Normal P–P plot of standardized residuals, $\{z_t\}$.

the theoretical normal distribution. In both figures, the standardized residuals are well matched with the theoretical normal distribution.

In sum, we estimated and removed a linear trend and monthly seasonal components from the original time series and a systematic pattern from the deseasonalized and detrended time series. The final residuals satisfy the residual assumptions: zero mean, constant variance, no influencing outliers, no autocorrelation, and the normality of distribution. We can now more confidently interpret our estimated model of the monthly violent crime rates and forecast future values based on the model.

## (3) FURTHER READING

Berry, W. D. 1993. *Understanding Regression Assumptions.* Sage University Paper Series on Quantitative Applications in the Social Sciences, no. 92. Newbury Park, CA: Sage.

Cromwell, J. B., W. C. Labys, and M. Terraza. 1994. *Univariate Tests for Time Series Models.* Sage University Paper Series on Quantitative Applications in the Social Sciences, no. 99. Newbury Park, CA: Sage.

Fox, J. 1991. *Regression Diagnostics.* Sage University Paper Series on Quantitative Applications in the Social Sciences, no. 79. Newbury Park, CA: Sage.

Lewis-Beck, M. 1980. *Applied Regression: An Introduction.* Sage University Paper Series on Quantitative Applications in the Social Sciences, no. 22. Newbury Park, CA: Sage.

Razali, N. M., and Y. B. Wah. 2011. "Power Comparisons of Shapiro-Wilk, Kolmogorov-Smirnov, Lilliefors and Anderson-Darling Tests." *Journal of Statistical Modeling and Analytics* 2:21–33.

Seier, E. 2002. "Comparison of Tests for Univariate Normality." *InterStat Statistical Journal* 1:1–17.

# FOUR

# *Forecasting*

IN THE SOCIAL SCIENCES, forecasting is of multiple uses. First, we employ time series analysis to scientifically forecast voting rates, election results, crime rates, stock prices, the proportion of citizens who favor a certain policy alternative, and so on. Second, we use forecasted values to evaluate the accuracy of an estimated time series model. We compare forecasted values with observed values to check how well they match. We report forecast accuracy measures along with an estimated model, as we report $R^2$ or percentage correctly classified in regression analysis of cross-sectional data. Third, we compare forecasted values with the observed cases after an intervention point to determine whether a policy has an intended impact (see chapter 7).

OLS regression analysis of cross-sectional data is concerned primarily with the relationship between the independent variables and the dependent variable. In univariate time series analysis, forecasting is as important as the estimation of the relationship. Sometimes, forecasting is the sole purpose of time series analysis. In this chapter, we discuss how to forecast values and how to measure the accuracy of forecasted values.

## (1) HOW TO FORECAST VALUES

In univariate time series analysis, we can forecast future values in two ways. First we figure out the systematic components of a time series, and then we forecast future values based on the estimated systematic components. If a fitted model consists only of a trend and white noise, forecasting will be the same as that in OLS regression analysis: we multiply a value of the independent variable, Time, by the estimated trend coefficient and then add the intercept.

If a fitted model contains seasonal components and/or systematically behaving time-dependent residuals, we forecast future values by repeating the model estimation procedure but in the reverse order (see chapter 2). When we model a time series, we first estimate and eliminate seasonal components, while keeping a trend and a systematic pattern in the time series. After we have removed estimated seasonal components from the time series, we estimate a trend. If a systematic pattern still exists in the deseasonalized and/or detrended series, we finally estimate the systematic pattern of residuals.

To forecast a lag-1 future value, we first forecast a lag-1 future residual based on the estimated systematic behavior of the observed residuals. We also forecast a lag-1 future point on the estimated trend line by multiplying a lag-1 future value of the independent variable, Time, by the estimated trend coefficient. Then, we sum up these two forecasted values and the intercept. When we subtracted a nonzero mean from the deseasonalized and/or detrended series, we add the mean back to the sum. Finally, we adjust the sum with the estimated seasonal component for a lag-1 time point by adding the component to the sum, if we estimated seasonal components with the additive method. We multiply the sum by the estimated seasonal component, if we estimated seasonal components with the multiplicative method. When we estimated seasonal components with the multiplicative method, multiplying the estimated seasonal component has to be the final step of the forecasting procedure. When we estimated seasonal components with the additive method, adding the estimated seasonal component does not have to be the final step.

Among these steps, forecasting future residuals is more complicated and time-consuming than the other steps, if we manually calculate them. To forecast a future value on the estimated trend line, we simply multiply an estimated trend and any future value of time. For example, we multiply 132 for the monthly violent crime rates to forecast a value at time $t + 12$. On the other hand, we forecast residuals with past residuals. Except for the lag-1 future residual, all the others are forecasted with forecasted residuals. Since we do not know the future residuals, we need to forecast them one by one. Therefore, we cannot forecast the residual at $t + i$ until we forecast the residual at $t + i - 1$. For example, to forecast the lag-3 future residual, we forecast the lag-1 future residual with the current residual for an AR(1) process, with the current and lag-1 previous residuals for an AR(2) process, . . . , and with the current to lag $p - 1$ previous residuals for an AR($p$) process. Then, we are ready to forecast the lag-2 future residual with the forecasted lag-1 future residual in the same way. Finally, we forecast the lag-3 future residual with the forecasted lag-1 and lag-2 future residuals. In reality, we determine how many future residuals will be forecasted, and then programs will report the forecasted residuals.

Forecasting based on an MA($q$)/ARMA($p, q$) model is generally less effective than that based on an AR($p$) model because an MA($q$)/ARMA($p, q$) model contains random components, $\{Z_t\}$, that are less stable than an autoregressive process. This is a reason why we usually estimate an MA process to account for a systematic pattern that remains after an autoregressive component was removed from the residuals in the social sciences.

Second, if forecasting is our sole purpose, we can skip the modeling procedure and use an algorithm to forecast future values. We can employ algorithms such as the simple exponential smoothing algorithm, Holt's linear trend algorithm, Brown's linear trend algorithm, damped trend algorithm, simple seasonal exponential smoothing algorithm, and Winter's additive or multiplicative exponential smoothing algorithm (Brockwell and Davis 2002; Chatfield 2004). All these algorithms are based on special types of exponential smoothing. With most statistical programs, we can forecast values with these algorithms. Understanding the mathematical characteristics of these algorithms is not required in most social science studies. We just need to know when to select each of these algorithms to forecast values.

To select one among these algorithms, we need to gather information on the time-dependent behavior of a time series by visually inspecting the time series (see chapter 2). Simple exponential smoothing is an appropriate technique when a time series does not contain a trend and seasonal components. Holt's linear trend or Brown's linear trend is an appropriate technique when a time series contains a linear trend but no seasonal components. Damped trend is an appropriate technique when a time series contains a damping or exponentially increasing trend, rather than a linear trend, and no seasonal components.[1] Simple seasonal exponential smoothing is an appropriate technique when a time series contains seasonal components that are constant over time, but no trend. Winter's additive or multiplicative exponential smoothing is an appropriate technique when a time series contains a linear trend and seasonal components that do not depend on the level of the series or that do depend on the level, respectively.

We may not be able to identify a time-dependent behavior with confidence through visual inspection alone. Therefore, we may apply several algorithms to an observed time series and then compare forecast errors, the difference between an observed time series and a forecasted series, $\{X_t\} - \{\hat{X}_t\}$, or other forecast accuracy measures, such as the average of white noise, MAE, MAPE and $R^2$ (see below). We can also check the randomness of residuals that are produced by each algorithm (see chapter 2). After selecting one that does better than the others, we can use it to forecast unobserved future values.

Forecasting with an algorithm may be more efficient in terms of mean squared errors than forecasting with an estimated model. The purpose of most social science studies, however, is not only to forecast future values but also to figure out the time-dependent behavior of observations. Even when forecasting is of primary concern, we may be required to explain how we arrived at our forecast. Thus, in most social science studies, we usually fit a model to a time series and then forecast with the fitted model, rather than simply forecasting with a particular type of algorithm. With this fitted model, we can provide the general movement of a statistically detectable long-term change, the regular short-term fluctuations with a period of less than one year, such as day, week, month, and quarter, and the systematic pattern of residuals as our explanations about how we arrived at our forecast.

In multiple time series analysis, forecasting is similar to that in OLS regression analysis. We multiply future values of the independent variable(s) in our model by their estimated coefficients and total them and an intercept. When we deseasonalize and whiten the dependent variable before we conduct multiple time series analysis, we adjust the forecasted values with the estimated seasonal components and systematic patterns of residuals. We may incorporate the estimation of seasonal components in a multiple time series model (see section 2B and table 10, chapter 6, page 140). In this case, we adjust forecasted values by adding the estimated coefficient of the month or quarter or week that we want to forecast. The forecasted value without the seasonal adjustment will be the forecasted value for the excluded category of the dummy variable scheme for seasonal components, that is, December in table 10.

## (2) MEASURING THE ACCURACY OF TIME SERIES MODELS

The accuracy of a time series model is measured by visually or statistically comparing forecasted values and observed values. In this respect, the accuracy of an estimated time series model is identical to the model's forecast accuracy. When a model perfectly forecasts future values, the difference between observed values and forecasted values will be zero.

To measure the accuracy of a time series model, we compare observed values and forecasted values in two slightly different ways. First, we compare forecasted values and observed values during a period that is withheld for validation. We can check visually how well observed values and forecasted values match during a period withheld for validation. We can also calculate a bivariate correlation between forecasted values and observed values during the period.

In the social sciences, however, we may not appropriately test the accuracy of an estimated model by visually or statistically comparing forecasted values and observed values during a period that is withheld for validation. Various factors may operate on a time series, and therefore trend and/or other systemic components may change. For example, if a strong crime-control policy is implemented, crime rates may decrease. In this case, even if observed values and forecasted values are significantly different during a withheld period, such deviation may not evidence the lack of fit of an estimated time series model. In interrupted time series analysis, a significant deviation between forecasted values and observed values after an intervention point is a vital part of the research hypothesis that the intervention has an expected impact (see chapter 7).

Second, we compare observed values and predicted values for the same period covered by our time series analysis. In this case, differences between an observed series and a predicted series, $\{X_t - \hat{X}_t\}$, are errors that are left over at the end of a time series modeling process.

There are various measures of accuracy that compare observed values and predicted values for the same period covered by our time series analysis. The smaller the difference between an observed time series and a predicted time series is, the more accurate a time series model is. Thus, the smaller these accuracy measures are, the more accurate our model is. An exception is the coefficient of determination ($R^2$), which is larger for a more accurate model. The most basic measure of forecast accuracy is an average of the errors:

$$\frac{\sum_{t=1}^{n}(X_t - \hat{X}_t)}{n}$$

where $n$ is the total number of cases.

A time series model whose average of the errors is closer to zero is more accurate. A weakness of this measure is that it can be small even when the difference between an observed time series and a predicted time series is large, if positive and negative deviations cancel each other. This measure is not appropriate as a measure of accuracy in univariate time series analysis. When we accurately model a time series, we estimate and then eliminate all systemic components (trend, seasonality, and AR($p$) and/or MA($q$) processes) from the time series and only white noise will be left over. Therefore, the difference between an observed time series and a predicted time series will be a realization of white noise, and the average of errors will be zero or at least negligible.

An alternative measure, in which positive and negative deviations do not cancel each other, is an average of the absolute errors (mean absolute error, MAE):

$$\frac{\sum_{t=1}^{n} |X_t - \hat{X}_t|}{n}$$

The smaller the MAE is, the more accurate a time series model is.

These two accuracy measures have a common weakness: an average will be proportional to the scale of individual values of a time series. Therefore, these accuracy measures should not be compared across time series datasets whose scales are different.

To overcome this weakness, we divide absolute errors by corresponding absolute observed values and then calculate their average. By multiplying by 100, we can express this average difference between an observed series and a predicted series as a percentage (mean absolute percentage error, MAPE):

$$\sum_{t=1}^{n} \frac{|X_t - \hat{X}_t|}{|X_t|} \times \frac{1}{n} \times 100$$

The MAPE has no upper bound, although it has zero as its lower bound. An MAPE equal to zero implies a perfect prediction. An MAPE above 100%, however, may not imply a zero accuracy. For predicted values that are too low, the percentage error cannot exceed 100%, but for predicted values that are too high, there is no upper limit to the percentage error. For example, if all predicted values are zero regardless of observed values, the MAPE will be 100. If all predicted values are exactly twice the corresponding observed values, the MAPE will be 100. If all predicted values are more than twice the corresponding observed values, the MAPE will continue to increase.[2]

The MAPE has several drawbacks. First, as usual in the statistics based on the mean, the MAPE is sensitive to extreme values. A few predicted values that are too high may substantially increase the MAPE. Second, the MAPE cannot be used when zero exists among observed values, because there will be division by zero. Third, the MAPE is scale-sensitive. When observed values are quite small, the MAPE will take on extreme values. For instance, when a predicted value is 1.2 and an observed value is 1, the MAPE is 20 (0.2 × 100). When a predicted value is 0.3 and an observed value is 0.1, the MAPE is 200 (2 × 100) for the same absolute amount of error, 0.2.

To overcome this weakness, we can divide the absolute difference between an observed value and a predicted value by the sum of the observed value and the predicted value. This measure is termed the symmetric mean absolute percentage error (SMAPE):

$$\sum_{t=1}^{n} \frac{|X_t - \hat{X}_t|}{(|X_t| + |\hat{X}_t|)/2} \times \frac{1}{n} \times 100$$

In the denominator, $|X_t| + |\hat{X}_t|$ is divided by 2 to make the SMAPE comparable to the MAPE, whose denominator is $X_t$. The SMAPE has zero as a lower bound and 200 as an upper bound. Sometimes the denominator, $(|X_t| + |\hat{X}_t|)/2$, is replaced by $|X_t| + |\hat{X}_t|$ to make the SMAPE range from 0 to 100, which can be more easily interpreted as a percentage.

As in the above formulas for the MAPE and the SMAPE, we need to take absolute values of the denominator as well as the numerator, if an original time series and a predicted time series include negative values. If we take absolute values of only the numerator, negative values in the original series will decrease the MAPE. In the SMAPE, positive and negative deviations between an observed series and a predicted series will cancel each other in the denominator, and therefore our estimation of the mean percentage error will be biased. When we calculate the SMAPE, it is important to take the absolute values of $X_t$ and $\hat{X}_t$ in the denominator, separately. When $X_t$ is positive, $\hat{X}_t$ can be negative, and vice versa. This can happen especially when we deal with residuals because their values are usually very small and therefore a small amount of prediction error can easily make the sign of predicted values different from that of original values. Therefore, if we take absolute values of $X_t + \hat{X}_t$, the denominator may become smaller or greater than it should be.

The SMAPE has several drawbacks. First, the SMAPE adds more weight to under-forecasts, $\hat{X}_t < X_t$, than to over-forecasts, $\hat{X}_t > X_t$, because $|X_t| + |\hat{X}_t|$ is greater with over-forecasts than with under-forecasts. Therefore, it may be more appropriate to call it the adjusted MAPE, as Armstrong (1985) originally termed it, rather than the symmetric MAPE. Second, interpretation of the SMAPE is less intuitive with regard to the forecast accuracy of an estimated model than interpretation of the MAPE. It is easier and more intuitive to compare the amount of error to the sum of the observed values, as in the MAPE, than to the sum of the observed values and the predicted values, as in the SMAPE. Third, values predicted by different models may be significantly different from each other, even when we deal with the same observed values. Therefore, when we compare different models, their SMAPE's denominators, $|X_t| + |\hat{X}_t|$, may be different from each other for the same observed value. For this reason, when we compare model fits of different models, the MAPE, whose denominator, $|X_t|$, remains unchanged across models, may be more appropriate than the SMAPE.

Another measure in which positive and negative deviations do not cancel each other is the mean squared error (MSE). This measure squares the difference between an observed time series and a predicted time series, adds up the squared deviations, and then divides the sum by the total number of cases:

$$\frac{\sum_{t=1}^{n}(X_t - \hat{X}_t)^2}{n}$$

The square root of MSE (RMSE) describes the difference between a time series and its predicted values in the same units as the time series.[3] It is similar to the most basic measure of accuracy, an average of the errors, but it may not be zero even for white noise.

Univariate time series analysis is identical to bivariate OLS regression analysis with Time as the independent variable. Therefore, we may also employ the coefficient of determination, $R^2$, or the coefficient of nondetermination, $1 - R^2$, to measure the accuracy of a time series model. The coefficient of determination is the ratio of the variation of the dependent variable, which is explained by the independent variable, to the total variation of the dependent variable. In univariate time series analysis, it is the ratio of the variation of a time series, which is explained by its estimated temporal patterns (trend, seasonal components, and an ARMA$(p, q)$ process), to the total variation of the time series:

$$\frac{\sum(\hat{X}_t - \bar{X}_t)^2}{\sum(X_t - \bar{X}_t)^2}$$

The coefficient of nondetermination, $1 - R^2$, is the ratio of the unexplained variation to the total variation:

$$\frac{\sum(X_t - \hat{X}_t)^2}{\sum(X_t - \bar{X}_t)^2}$$

In time series analysis, we estimate and remove systematic components from residuals. The unexplained variation will be smaller in time series analysis than in OLS regression analysis of cross-sectional data, in which residuals are left unexplained. Therefore, the coefficient of determination will be very large, approaching unity, if a time series is accurately modeled.

We need to replace $R^2$ with adjusted $R^2$ in multiple time series analysis, as in multiple OLS regression analysis. This measure includes a penalty factor to compensate for the increase of the coefficient of determination due to the increase of the number of independent variables (Neter, Wasserman, and Kutner 1990).

When calculating accuracy measures, we need to determine whether we will measure the accuracy of an entire model, which includes trend, seasonality, and systematic patterns of residuals, or of a part of the model, such as an estimated systematic pattern of residuals. As discussed in chapter 2, we may bypass the estimation of trend and/or seasonality and directly estimate a systematic pattern of residuals. To the contrary, we may damp residual fluctuations to estimate trends. Therefore, we may not always be interested in the accuracy of an entire model. Accuracy measures of an entire time series model may be significantly different from those of a part of the model.

When our time series model is accurate, both the difference between observed values and values predicted by an entire model, $\{X_t - \hat{X}_t\}$, and the difference between observed values of a deseasonalized and/or detrended series and values of residuals predicted by an estimated systematic pattern,

$\{Y_t - \hat{Y}_t\}$, will be white noise, $\{Z_t\}$. Therefore, whether we estimate the accuracy of an entire model or only of an estimated systematic pattern of residuals, statistics whose denominator is the number of cases, such as the MAE, the MSE, and the RMSE, will be identical.

However, if trend and/or seasonality significantly explain the variation of $\{X_t\}$, statistics whose denominator includes observed values may change according to whether we estimate the fit of an entire model or of an estimated systematic pattern of residuals. The denominator in the latter—$Y_t$ or $(Y_t + \hat{Y}_t)$ or $(Y_t - \bar{Y}_t)$—will be smaller than the denominator in the former—$X_t$ or $(X_t + \hat{X}_t)$ or $(X_t - \bar{X}_t)$—if trend and/or seasonality significantly explain the variation of $\{X_t\}$. For instance, the MAPE, the SMAPE, and $R^2$ will be greater when observed and predicted values of residuals, $Y_t$ and $\hat{Y}_t$, are compared with each other than when observed and predicted values of an original time series, $X_t$ and $\hat{X}_t$, are compared with each other.

In the coefficient of determination, the numerator is also likely to be different according to whether we measure the accuracy of an identified systematic pattern of residuals or of an entire time series model. The numerator is $\hat{Y}_t - \bar{Y}_t$ in the former and $\hat{X}_t - \bar{X}_t$ in the latter. Absolute values of $\hat{Y}_t$ are likely to be smaller than those of $\hat{X}_t$, if trend and/or seasonality are statistically significant. In addition, $\bar{Y}_t$ will be either zero or negligible, while $\bar{X}_t$ may be significantly different from zero.

If we begin our modeling process with an original time series that contains trend and seasonality as well as systematically behaving residuals, statistical programs will report model fits of an entire time series model. In this case, we can refer to the reported MAE, MSE, and RMSE to determine the accuracy of an estimated residual pattern. However, we need to recalculate the MAPE, SMAPE, and $R^2$. We may begin a new modeling process with residuals and let programs report these statistics. We can calculate the MAPE by calculating and saving the absolute values of white noise (absolute error), and then dividing the saved values by the absolute values of $Y_t$ and saving them, and then letting statistical programs report the mean of the saved values. For the SMAPE, we additionally calculate $|Y_t| + |\hat{Y}_t|$, divide the absolute errors by $|Y_t| + |\hat{Y}_t|$ and save them, and finally let statistical programs report the mean of the saved values. The coefficient of determination or the coefficient of nondetermination can be calculated by replacing $X_t$, $\hat{X}_t$, and $\bar{X}_t$ with $Y_t$, $\hat{Y}_t$, and $\bar{Y}_t$, respectively, in the above formula.

We generally expect the level of accuracy to be much higher in univariate time series analysis than in multiple time series analysis (for multiple time series analysis, see chapter 6) or in OLS regression analysis of cross-sectional data. In univariate time series analysis, we attempt to explain the temporal behavior of the dependent variable itself, rather than the relationship between selected explanatory variables and the dependent variable. The behavior of the dependent variable will reflect the influence of all important factors on the dependent variable. The level of accuracy in univariate time series analysis will depend on the characteristics of a time series. For instance, we expect our model to be more accurate when the dependent variable tends to be highly stable than when it is unstable.

In OLS regression analysis of cross-sectional data and in multiple time series analysis, we explain the dependent variable with one or more explanatory variables. No one expects selected explanatory variables to perfectly explain the dependent variable in the social sciences. Explanatory variables in our model may explain only a small amount of variation of the dependent variable. However, if the variables and the equation are statistically significant, we take them as evidence for our theory about the relationship between the explanatory variables and the dependent variable.

## (3) THE CASE OF MONTHLY VIOLENT CRIME RATES, 1983–1992

In the modeling process of the monthly violent crime rates from 1983 to 1992, we figured out seasonal components of lag 12, a linear trend, and an AR(2) process. Now we are ready to forecast future monthly violent crime rates. Since the monthly violent crime rates from 1983 to 1992 have a seasonality of period 12, we will forecast the next 12 units, that is the monthly violent crime rates from January to December 1993. Table 7 presents the 12 crime rates forecasted by the estimated model in chapter 2.

The monthly violent crime rates have 120 observations. Therefore, values of the independent variable, Time, which we will input to forecast the next 12 units, will range from 121 (January 1993) to 132 (December 1993). To forecast future crime rates, we repeat the model estimation procedure in chapter 2 in the reverse order. We first forecast residuals based on the estimated AR(2) model (first column). Second, we forecast future crime rates based on the estimated trend alone (second column). Third, we total these two forecasted values and an intercept to forecast future crime rates without seasonal variation. Fourth, we add the corresponding seasonal components (see table 1,

TABLE 7. Forecasted Monthly Violent Crime Rates for 1993

| Case | Forecasted residuals, $\{\hat{Y}_j\}$[a] | Trend × Time | Forecasted crime rate[b] | Forecasted crime rate[c] |
|------|------|------|------|------|
| 121 | – 2.42 | 22.81 | 59.63 | 60.17 |
| 122 | – 2.15 | 23.0 | 55.88 | 56.04 |
| 123 | – 1.84 | 23.19 | 61.15 | 60.75 |
| 124 | – 1.60 | 23.37 | 61.04 | 60.31 |
| 125 | – 1.38 | 23.56 | 65.04 | 63.88 |
| 126 | – 1.19 | 23.75 | 66.31 | 64.8 |
| 127 | – 1.02 | 23.94 | 70.27 | 68.75 |
| 128 | – 0.88 | 24.13 | 71.07 | 69.16 |
| 129 | – 0.76 | 24.32 | 67.67 | 65.64 |
| 130 | – 0.65 | 24.51 | 67.78 | 65.69 |
| 131 | – 0.55 | 24.69 | 63.59 | 61.21 |
| 132 | – 0.47 | 24.88 | 65.32 | 62.83 |

[a] Mean corrected value (–0.05) was added.
[b] Intercept (41.961) was added.
[c] Crime rates forecasted by Winter's additive exponential smoothing algorithm.

chapter 2, page 26) to these forecasted crime rates (third column). As explained above, we do not have to follow this procedure in order if we estimated seasonal components with the additive method. If we estimated seasonal components with the multiplicative method, however, we have to multiply them after we complete the first three steps.

According to the forecasted monthly violent crime rates (third column), crime rates are expected to increase by about 1.5% on average in 1993 over 1992, along with monthly variations that are lowest in February (lower by about 10.73% than the average violent crime rate in 1993) and highest in August (higher by about 9.08% than the year's average violent crime rate).

Actual crime rates tended to be lower than expected in 1993. Ten out of the twelve forecasted crime rates are higher than the corresponding actual crime rates. The difference between forecasted and observed crime rates ranged from –1.08 (December) to 5.89 (February) crimes per 100,000 inhabitants, with the mean of the twelve differences equal to 2.33 crimes per 100,000 inhabitants. Paired-samples $t$ test $(3.5, p = .005)$ indicates that crime rates were significantly lower than we could expect according to the estimated systemic patterns of crime rates in 1983–1992.

Although the crime rates observed in 1993 were significantly lower than expected, the difference between the observed and forecasted crime rates may

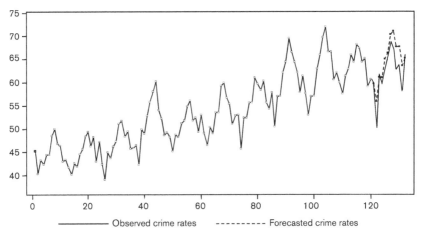

FIGURE 35. Observed monthly violent crime rates from 1983 to 1993 and twelve rates forecasted for 1993.

not be evidence against the accuracy of our estimated model. According to the visual comparison of the twelve observations in 1993 and the twelve forecasted crime rates, the estimated model appears to be accurate. Figure 35 plots the twelve crime rates forecasted by the estimated model, along with the 132 observed crime rates. Figure 36 plots the twelve crime rates forecasted by the estimated model, along with the twelve observed crime rates in 1993. According to these figures, the forecasted values and the observed values are well matched. The forecasted directions of change are in accord with the observed directions, except for August. In addition, the relative amount of change of the forecasted monthly crime rates is similar to that of the observed crime rates. The bivariate correlation between the twelve forecasted values and the corresponding observed values is 0.89 ($p$ = .000).

As discussed above, trend and/or other systematic temporal components may continually change under the influence of various factors. Therefore, we may test the accuracy of an estimated model more appropriately by comparing observed values and predicted values during the same period covered by time series analysis than during a period withheld for validation.

A linear trend and seasonal components of lag 12 appear to be obvious in the monthly violent crime rates from 1983 to 1992 (see chapter 2). Therefore, we do not have to compare several competing models, for example with linear or curvilinear trend and with monthly or no seasonal variations. If we cannot clearly determine the functional forms of trend and seasonality, we can compare the accuracy of these competing models.

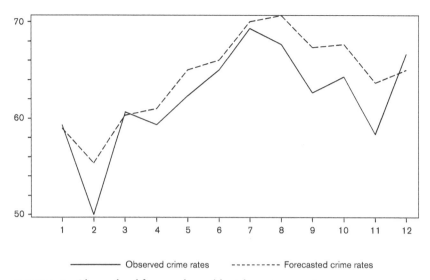

FIGURE 36. Observed and forecasted monthly violent crime rates in 1993.

We have already found that an AR(2) process represents the systematic pattern of residuals, $\{Y_t\}$, better than AR(1) and ARMA(1, 1) (see chapter 2). Thus, we do not have to compare these three processes again at this stage of our time series modeling procedure. Here, we simply double-check whether AR(2) is better than the other two processes in terms of the model accuracy.

When an AR(2) process is selected for the stochastic process of the deseasonalized and detrended crime rates, the MAE of $\{Z_t\}$ is 1.13. The RMSE is 1.42. When AR(1) is selected, these measures of accuracy are 1.142 and 1.48, respectively. When ARMA(1, 1) is selected, they are 1.141 and 1.43, respectively. The MAE and the RMSE tell us that AR(2) represents the systematic pattern of $\{Y_t\}$ better than AR(1) and ARMA(1, 1).

When AR(2) is selected for the stochastic process of the deseasonalized and detrended crime rates, the MAPE and the SMAPE are 394.96 and 99.74 (49.87, if the denominator is not divided by two), respectively. When AR(1) is selected, the two measures of accuracy are 447.20 and 98.64 (49.32), respectively. When ARMA(1, 1) is selected, the two measures of accuracy are 417.21 and 99.56 (49.78), respectively.

According to the MAPE, which is more appropriate than the SMAPE when we compare the accuracy of different models (see above), AR(2) appears to represent the deseasonalized and detrended series better than the other two processes. The SMAPE is highest for AR(2). However, the differences among the three SMAPEs are negligible, especially if we compare them to the differences among three MAPEs.

When the absolute error $|X_t - \hat{X}_t|$ $(= |Y_t - \hat{Y}_t| = |Z_t|)$ is compared to the values of the original series, $\{X_t\}$, the MAPE and the SMAPE are 2.15 and 2.14 (1.07, if the denominator is not divided by two), respectively, for the AR(2) model. When AR(1) is selected for the systematic pattern of the deseasonalized and detrended crime rates, the two measures of accuracy are 2.161 and 2.164 (1.082), respectively. When ARMA(1, 1) is selected, the two measures of accuracy are 2.161 and 2.163 (1.082), respectively. Like the MAE and the RMSE, the MAPE and the SMAPE show that AR(2) represents the systematic pattern of $\{Y_t\}$ better than AR(1) and ARMA(1, 1).

The coefficient of determination for the full model that includes AR(2) is .968. It is .965 for the full model that includes AR(1) and .967 for the full model that includes ARMA(1, 1). When the explained residual variance is compared to the total variance of $\{Y_t\}$, the coefficient of determination is .52 for the AR(2) model, .47 for the AR(1) model, and .50 for the ARMA(1, 1) model. All these statistics, along with the visual comparison (figure 36) and the bivariate correlation between the twelve forecasted values and the corresponding observations, confirm the goodness of fit of the estimated model.

As explained above, when we just want to forecast future values without explaining the systematic patterns of the monthly violent crime rates, we can employ an algorithm. The monthly violent crime rates from 1983 to 1992 consist of a linear trend, seasonal components of lag 12, and systematically patterned residuals. Therefore, Winter's additive exponential smoothing is appropriate to forecast future crime rates. The forecasted crime rates for the twelve months of 1993 are reported in table 7 (fourth column). Only a slight difference exists between these forecasted values and the values forecasted with the estimated model. The bivariate correlation between these two groups of forecasted values is .99 ($p = .000$). Eight out of the twelve crime rates forecasted by Winter's algorithm are higher than the corresponding actual crime rates. The forecasted directions of change of crime rates are in accord with the observed directions, except for August and October. The bivariate correlation between the forecasted values and the corresponding observed values is .85 ($p = .000$). The MAE is 1.17, and the RMSE is 1.42.

As in this case, whether we forecast future values with the estimated model or with an appropriate algorithm, the results are almost identical, with a few minor differences. We prefer the forecast with an estimated model because we are usually expected to provide explanation of our forecast in the social sciences.


## (4) FURTHER READING

Armstrong, J. S. 1985. *Long-Range Forecasting: From Crystal Ball to Computer.* New York: Wiley.

————. 2001. *Principles of Forecasting: A Handbook for Researchers and Practitioners.* Norwell, MA: Kluwer Academic.

Armstrong, J. S., and F. Collopy. 1992. "Error Measures for Generalizing about Forecasting Methods: Empirical Comparisons." *International Journal of Forecasting* 8:69–80.

Diebold, F. X., and R. Mariano. 1995. "Comparing Predictive Accuracy." *Journal of Business and Economic Statistics* 13:253–263.

Fair, R. C. 1980. "Evaluating the Predictive Accuracy of Econometric Models." *International Economic Review* 21:355–378.

Gardner, E. S., Jr., and E. McKenzie. 1985. "Forecasting Trends in Time Series." *Management Science* 31:1237–46.

Granger, C. W. J., and P. Newbold. 1986. *Forecasting Economic Time Series,* 2nd ed. Orlando: Academic Press.

Rescher, N. 1998. *Predicting the Future: An Introduction to the Theory of Forecasting.* Albany: SUNY Press.

# FIVE

# *Smoothing*

UNTIL NOW, WE HAVE DISCUSSED how to decompose a time series into a trend, seasonal components, systematically patterned residuals, and white noise; how to diagnose the appropriateness of an estimated time series model; how to forecast values with an estimated model; and how to measure the accuracy of an estimated model. This is the typical time series analysis procedure in the social sciences.

Sometimes, we may not need to identify all the systematic patterns in a time series. We may be concerned about its trend, a long-term behavior of the time series, but not about other systematic patterns—seasonality or an ARMA($p$, $q$) process in residuals—that cause short-term fluctuations of the time series. In this case, we may not need to estimate and then eliminate the short-term fluctuations. Instead, we may simply reduce the short-term fluctuations in the time series to make the trend stand out more clearly. For example, Green, Gerber, and De Boef (1999) show that constructing a weighted average of party identification based on the ratio of signal to noise in each poll helps us reduce the random movements produced by sampling errors in public opinion surveys repeated over time and identify genuine movements in party identification more accurately.

To reduce short-term fluctuations in a time series without estimating them, and thereby to make its long-term trend stand out more clearly, we can employ a technique called *smoothing*. We smooth a time series by applying a filter to the time series. There are various types of smoothing, among which simple *moving average smoothing* and *exponential moving average smoothing* are frequently employed.

## (1) MOVING AVERAGE SMOOTHING

Moving average smoothing uses a moving average of a particular order as a filter to reduce short-term fluctuations. Two types of moving average smoothing are usually employed. One is *prior moving average smoothing*. The other is *centered moving average smoothing*. In prior moving average smoothing, a moving average of order $q$ is the mean of $q$ consecutive observations preceding the current observation in a time series. For example, a moving average of order 6 consists of means of every six consecutive observations that move over time.

In centered moving average smoothing, a moving average of order $q$ is also the mean of $q$ observations in a time series, but the selection of observations is different from prior moving average smoothing. Instead of using only the past observations, the $q$ observations are equally divided before and after the point where the mean is calculated. That is, we select the first $(q - 1)/2$ observations before the current observation; the current observation; and $(q - 1)/2$ observations after the current observation.

Centered moving average usually has an odd number as its order. If $q$ is even, selection is not symmetric. For example, if $q$ is 4, we select one observation before the current observation, the current observation, and two observations after the current observation. In this case, the points where the mean is calculated, such as 2.5, 3.5, and so on, do not exist. To make this even-order moving average symmetric, we may apply a moving average smoothing of order 2 to the values smoothed by a moving average of order 4. That is, we average every two smoothed values. Then the points where the mean is calculated will be 3, 4, 5, and so on. If $q$ is 12, the points where the mean is calculated will be 7, 8, 9, and so on.

For prior moving average smoothing of order $q$, a moving average cannot be calculated for the first $q$ observations, and therefore the number of cases decreases by $q$. For centered moving average smoothing of order $q$, the number of cases decreases by $q - 1$, if $q$ is odd, because a moving average cannot be calculated for the first $(q - 1)/2$ observations and the last $(q - 1)/2$ observations. If $q$ is even and if we apply a moving average smoothing of order 2 to the smoothed values, the number of cases decreases by $q$ because a moving average cannot be calculated for the first $q/2$ observations and the last $q/2$ observations.

Averaging multiple consecutive observations in a time series will reduce short-term fluctuations and make a long-term trend stand out. If a time series consists of a linear trend and systematically patterned residuals, a moving average filter of order $q$ will reduce the residual variance by making fluctuating residuals cancel each other. If a time series contains seasonal components, a moving average filter whose order ($q$) is the same as the length of the seasonality will reduce the variance of the seasonal components in the time series. For example, a moving average of order 4 will translate every four quarterly observations into an averaged value, thereby reducing the seasonal variation in the time series.

In a new time series which was transformed by a moving average filter, a trend will stand out more clearly, as long as the trend remains fixed throughout the observation period. If a trend is nonlinear and changes significantly during the observation period, we need to smooth a time series with caution. A smaller order of smoothing is generally more appropriate when a trend significantly changes during the observation period. When we smooth a time series with a moving average filter of large order, we may have a distorted trend estimate. As the order of smoothing increases, the number of observations that are included in the averaging process increases. Consequently, the chance increases that smoothing will average out not only short-term fluctuations of residuals but also a long-term nonlinear trend.

When a trend obviously changes during the observation period, we may need to divide the observation period based on our theoretical consideration of the behavior of the trend and then analyze the parts separately. For example, in an interrupted time series analysis in which we usually expect the impact of a program to change a trend before and after the implementation of the program, we need to divide the observation period into the parts before and after the treatment point. Then we separately smooth each of the two divided time series to estimate their trends (see chapter 7).

The selection of an appropriate order of smoothing will require trial and error, along with theoretical consideration and visual inspection of a smoothed time series. Gaining a rough idea of an appropriate type of a long-term process of a time series, such as a linear or curvilinear trend, may also help us select an appropriate order of smoothing. In addition, it may be helpful to check residuals after we estimate and then eliminate a trend from a smoothed time series. Generally, the smaller the variance of the residuals, the more appropriate the order of smoothing.

## (2) EXPONENTIAL SMOOTHING

In prior or centered moving average smoothing, every observation included in the calculation of averages is given equal weight. That is, moving average smoothing does not take into account that temporally more distant observations tend to have less influence on the current observation. To consider this possibility, we may give less weight to the more distant cases, for example, to the first and last cases in every $q$ observations in centered moving average smoothing. That is, we may not divide each observation equally by $q$. Instead, we may divide the first and last cases in every $q$ observations by $2q$ and the remaining cases by $q$. Even in this case, all the observations between the first and last ones are given equal weight.

*Exponential smoothing* or *exponential moving average smoothing* is different from moving average smoothing in the following two ways. First, exponential smoothing applies exponentially decreasing weights to previous moving averages. Second, exponential smoothing considers all past observations (see below), placing a greater weight on a more recent observations. On the other hand, moving average smoothing considers only $q$ observations at a time to calculate averages, treating all or most of the $q$ observations equally.

Based on the assumption that the influence of past observations decreases exponentially over time, exponential smoothing applies a certain ratio $\alpha$ (between 0 and 1), which is termed a *smoothing constant*, to a current observation, and a *smoothing weight*, $1 - \alpha$, to a lag-1 previous moving average (Brown 1963):

$$MA_t = \alpha X_t + (1 - \alpha)MA_{t-1}$$

This equation describes that the current moving average is equal to the sum of an $\alpha$ portion of the current observation, $X_t$, and a $1 - \alpha$ portion of the lag-1 moving average. For example, if $\alpha$ is .2, the current moving average will be equal to the sum of one-fifth of the current observation and four-fifths of the lag-1 moving average.

The above equation is recursive and can be expressed as follows:

$$
\begin{aligned}
MA_t &= \alpha X_t + (1 - \alpha)MA_{t-1} \\
&= \alpha X_t + (1 - \alpha)[\alpha X_{t-1} + (1 - \alpha)MA_{t-2}] \\
&= \alpha X_t + \alpha(1 - \alpha)X_{t-1} + (1 - \alpha)^2 MA_{t-2} \\
&= \alpha X_t + \alpha(1 - \alpha)X_{t-1} + (1 - \alpha)^2[\alpha X_{t-2} + (1 - \alpha)MA_{t-3}] \\
&= \alpha X_t + \alpha(1 - \alpha)X_{t-1} + \alpha(1 - \alpha)^2 X_{t-2} + (1 - \alpha)^3 MA_{t-3} \\
&\quad\quad\quad\quad\quad \cdot \\
&\quad\quad\quad\quad\quad \cdot \\
&\quad\quad\quad\quad\quad \cdot \\
&= \alpha X_t + \alpha(1 - \alpha)X_{t-1} + \alpha(1 - \alpha)^2 X_{t-2} + \alpha(1 - \alpha)^3 X_{t-3} + \ldots + \\
&\quad \alpha(1 - \alpha)^{n-2} X_2 + (1 - \alpha)^{n-1} X_1
\end{aligned}
$$

In most time series analysis, we use data collected in the past. Therefore, $X_t$ will be the most recent observation, rather than the current observation, while $X_1$ is the last observation. The exponent of a smoothing weight begins with 1 for the lag-1 observation and ends with $n - 1$ for the last observation, where $n$ is the total number of observations.

As in this equation, a smoothing weight is not applied to the current observation. The smoothing weight for the current observation is 1, that is, $(1 - \alpha)^0$. For the current observation, there will be no decrease of influence, and therefore we do not have to weight its influence. On the other hand, a smoothing constant is not applied to the last observation. The last part of the above equation—$(1 - \alpha)^{n-1}X_1$—is the lag-1 moving average part of the second-to-last observation. Exponential smoothing does not include the moving average terms for the last observation, $X_1$—an $\alpha(1 - \alpha)^{n-1}$ portion of $X_1$, and a $1 - \alpha$ portion of its lag-1 moving average— because its lag-1 moving average does not exist in our time series data.

Since $\alpha$ ranges between 0 and 1, as $\alpha$ decreases, $(1 - \alpha)^k$ increases. Therefore, a smaller $\alpha$ gives a greater weight to the observations in the more distant past. With a larger $\alpha$, less weight is given to observations in the more distant past and the influence of the past observations on the current moving average will decrease faster. Therefore, with a smaller $\alpha$, we smooth a time series to a greater extent.

Selecting an appropriate value of the smoothing constant and the smoothing weight is based on subjective judgment. Generally, we may increase the smoothing constant as a time series shows a smaller degree of fluctuations. Just like selecting an appropriate order of moving average smoothing, selecting an appropriate smoothing constant will require trial and error. We may compare the goodness-of-fit of trend estimates, the variance of residuals, or the randomness of residuals after we fit a model to a time series smoothed with various smoothing constants.

## (3) THE CASE OF MONTHLY VIOLENT CRIME RATES, 1983–1992

As observed in chapter 2, the monthly violent crime rates from 1983 to 1992 contain seasonality of lag 12 and an autoregressive process of order 2, as well as random fluctuations that cannot be explained. We can see that there is an upward trend, but it is not possible to clearly identify the trend from the original time series (see figure 7, page 18). If our purpose is to check whether violent crime rates tend to increase significantly, and if we are not concerned about seasonal and other short-term fluctuations, we can smooth the crime rates and make the trend stand out more clearly.

As explained above, simple moving average smoothing reduces short-term fluctuations to a greater extent as its order increases because the number of observations included in the averaging process increases. Figure 37 shows monthly violent crime rates adjusted with moving average smoothing of order 3 and 12, along with the original crime rates. The moving average smoothing of order 3 does not satisfactorily reduce the short-term fluctuations. The moving average smoothing of order 12 significantly reduces the short-term fluctuations. In addition, it virtually eliminates the seasonal fluctuations. We can now clearly see that there is an upward linear trend. Since a prior moving average smoothing filter is applied, the first three and twelve cases of the original series have the system-missing values, respectively.

The trend estimates are very close to each other; the three trend lines are almost parallel in figure 37. The estimated trend is 0.197 for the crime rates smoothed with a moving average of order 3 and 0.201 for the crime rates smoothed with a moving average of order 12. The estimated trend for the original deseasonalized crime rates is 0.189 (see chapter 2). Given that the original crime rates range from 39.07 to 72.02, however, the difference, ranging from 0.004 to 0.012, is negligible, and the discussion of relative accuracy among these three estimates seems to be meaningless.

Figure 38 shows the monthly violent crime rates adjusted with exponential moving average smoothing with a smoothing constant of .1 and .3, along with the original crime rates. As explained above, exponential smoothing is calculated with all the monthly violent crime rates.

In Figure 38, we observe long stretches of plots on the same side above or below the expected trend line of the original crime rates, which are caused mostly by seasonality and systematically patterned residual fluctuations. A small smoothing constant is selected because we may need to give a greater weight to more distant observations to damp these fluctuations. When the smoothing constant is .1, seasonal fluctuations more clearly disappear, compared to when it is .3. A linear trend stands out more clearly in both adjusted series than in the original series. The estimated trend is 0.177 with smoothing

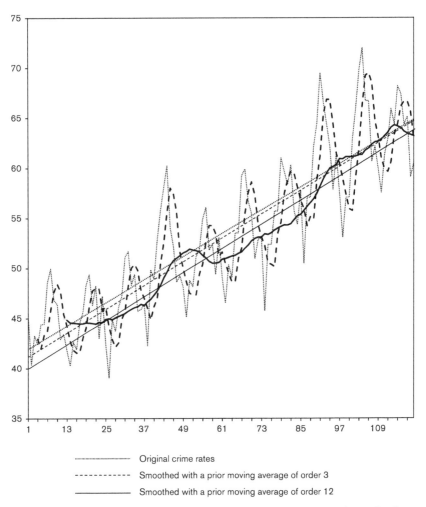

Original crime rates
----------- Smoothed with a prior moving average of order 3
——————— Smoothed with a prior moving average of order 12

FIGURE 37. Monthly violent crime rates: prior moving average smoothing of order 3 and 12.

constant .1 and 0.191 with smoothing constant .3. These estimates are slightly smaller than the estimated trends of the crime rates adjusted with a simple moving average smoothing.

In chapter 2, we estimated an AR(2) process and seasonal components of lag 12. We can eliminate the estimated AR(2) process from the deseasonalized series. That is, we can remove all the systematic patterns other than the trend from the original crime rates. In this case, the estimated trend is 0.188.

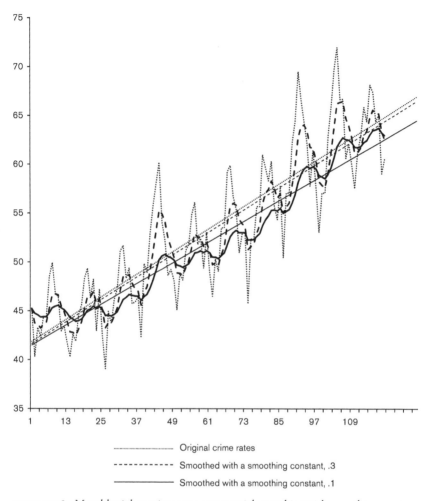

·············· Original crime rates

---------- Smoothed with a smoothing constant, .3

———— Smoothed with a smoothing constant, .1

FIGURE 38. Monthly violent crime rates: exponential smoothing with smoothing constant .1 and .3.

This is an estimate that is not biased by systematic fluctuations in the crime rates. Among the four estimated trends of the crime rates smoothed either with a moving average filter or with an exponential smoothing filter, the trend estimated with the exponential smoothing filter with a smoothing constant of .3 is closest to the estimated trend of the original crime rates. However, the differences among the estimated trends are not big enough to draw attention.

Brown, R. G. 1963. *Smoothing, Forecasting, and Prediction.* Englewood Cliffs, NJ: Prentice Hall.

Holt, C. C. 2004. "Forecasting Seasonals and Trends by Exponentially Weighted Moving Averages." *International Journal of Forecasting* 20:5–10.

Gardner, E. S., Jr. 1985. "Exponential Smoothing: The State of the Art." *Journal of Forecasting* 4:1–28.

———. 2006. "Exponential Smoothing: The State of the Art, Part II." *International Journal of Forecasting* 22:637–666.

Green, D. P., A. S. Gerber, and S. L. De Boef. 1999. "Tracking Opinion over Time: A Method for Reducing Sampling Error." *Public Opinion Quarterly* 63:178–192.

Hyndman, R. J., A. B. Koehler, J. K. Ord, and R. D. Snyder. 2008. *Forecasting with Exponential Smoothing: The State Space Approach.* Berlin: Springer.

Ord, J. K., A. B. Koehler, and R. D. Snyder. 1997. "Estimation and Prediction for a Class of Dynamic Nonlinear Statistical Models." *Journal of the American Statistical Association* 92:1621–1629.

# SIX

# Time Series Analysis with Two or More Time Series

## (1) CORRELATION AND REGRESSION ANALYSIS

### A. Multiple Time Series Analysis

In the previous chapters, we focused on univariate time series analysis. There are other types of time series analysis, such as bivariate, multivariate, and multiple time series analysis. These are confusing terminologies. As described in the previous chapters, univariate time series analysis explores time-dependent behaviors of a single time series. Thus, univariate time series analysis is bivariate OLS regression analysis in which a time series is the dependent variable and Time is the independent variable. *Bivariate time series analysis* explores the relationship between two time series. It is also bivariate OLS regression analysis, but the independent variable is a time series. *Multiple time series analysis* involves one dependent variable and two or more independent variables. Multiple time series analysis is different from multivariate time series analysis. Multiple time series analysis is similar to multiple regression analysis of cross-sectional data, exploring causal flows from two or more independent time series to one dependent time series. On the other hand, *multivariate time series analysis* involves two or more dependent variables at the current lag. If multivariate time series analysis includes two or more independent variables, we term it multivariate multiple time series analysis.

We focus on multiple time series analysis in this chapter. As in cross-sectional analysis, multiple (or bivariate) time series analysis is more frequently employed than multivariate time series analysis in the social sciences. In the social sciences, we are not familiar with the idea that multiple variables can be dependent on each other simultaneously. We usually select the dependent variable according to our theory. Then we explain the temporal behavior of the dependent variable in terms of its relationship with one or more independent variables, which are also selected according to our theory.

The formula for the least squares estimate in bivariate time series analysis is identical to that in bivariate regression analysis of cross-sectional data. The difference between the two estimates is that the dependent variable, $\{Y_t\}$, and the independent variable, $\{X_t\}$, are collected across time in bivariate time series analysis:[1]

$$\frac{\sum_{t=1}^{n}\left\{(Y_t - E(Y_t))\times(X_t - E(X_t))\right\}}{\sum_{t=1}^{n}(X_t - E(X_t))^2}$$

Multiple time series analysis is similar to multiple regression analysis of cross-sectional data. The least squares estimator for individual independent variable, $\{X_t\}$, in multiple time series analysis includes the $X$ matrix, the $Y$ matrix, and the inverse of the $X$ matrix, as in multiple regression analysis of cross-sectional data (Neter, Wasserman, and Kutner 1990, 225–270). In multiple regression analysis of cross-sectional data, we examine how the dependent variable changes as an independent variable varies, while the other independent variables are held constant. In multiple time series analysis, we examine how the dependent variable's temporal pattern changes when an independent variable temporally varies, while the other independent variables are held constant.

Several important differences exist between multiple (or bivariate) time series analysis and multiple (or bivariate) regression analysis of cross-sectional data. First, different from multiple regression analysis, the idea of multivariate relationship can be incorporated into multiple time series analysis. We assume a clear causal flow from one or more independent variables to one dependent variable in multiple regression analysis. If the independent variables are not independent of the dependent variable, our model has a serious specification error.

We also assume causality from one or more independent time series to one dependent time series in multiple time series analysis. As in multiple regression analysis of cross-sectional data, one variable cannot be both dependent on and independent of another variable at the current lag. In time series analysis, however, variables are collected across time through repeated regular observations, and the dependent variable at lag $k$ ($k > 0$) can influence the independent variable at the current lag. For example, there may be a causal flow from the size of the crime-control budget to violent crime rates, when both of them are observed across countries at the same point in time. To establish clear causality between these two variables, we may use the crime-control budget in the previous year. In time series analysis, the causal relationship between these two variables can be bidirectional. The larger crime-control budget may reduce the violent crime rate, which in turn may reduce the crime-control budget in the following year. Thus, although the violent crime rate is the dependent variable at the current lag, the violent crime rate at lag $k$ ($k > 0$) can be an independent variable that influences the crime-control budget at the current lag.[2]

Second, when we include a lagged dependent variable on the right-hand side of the equation to control the continuity of the dependent variable (see section 3), we deal with the autocorrelation problem differently in multiple time series analysis and in multiple regression analysis. With a lagged dependent variable on the right-hand side, autocorrelation between the dependent variable and the lagged dependent variable on the right-hand side may be statistically significant. In this case, autocorrelation will be reflected in the error term, and thereby an important regression assumption is violated. In multiple time series analysis, we prewhiten the dependent variable and thereby eliminate the possibility that the dependent variable and the lagged dependent variable are correlated with each other through the autocorrelation. In multiple regression analysis, we replace the lagged dependent variable with an instrumental variable. Specifically, we regress the lagged dependent variable on the exogenous factors, which can influence the dependent variable but are not included in the model. Then we use the predicted values for the instrumental variable. If the selected factors explain a significant amount of variance of the lagged dependent variable, the correlation between the instrumental variable and the residuals will be significantly reduced. For example, in a multiple regression model that explains stable party identification, we need to control the continuity of party identification by including lagged party identification on the right-hand side of the equation. In this case, we select factors, such as age, gender, level of education, level of family income, and city size, which are expected to influence party identification but are not included in the model. Lagged party identification is regressed on the selected factors. Then we use predicted values as the lagged party identification (see e.g. Fiorina 1982; Lockerbie 1989, 2002; Markus and Converse 1979; Shin 2012).

Third, in multiple time series analysis, we can include a lagged independent variable, along with the independent variable at the current lag, to properly take the dynamic of the variable's systematic time-dependent behavior into account. With this model, we can specify the relationship between the dependent variable and the independent variable, which may not be limited to the current lag but spread over a certain period. For example, presidential popularity may not only instantaneously adjust to current economic conditions, but also lag past economic conditions, because voters are likely to base their support of the president not only on current economic conditions but also on their memory of economic conditions in the last year (Beck 1991). As explained below, cross-correlations between the dependent variable and the independent variable help us determine the number of lags to be considered for the lagged values of the independent variable.

Fourth, we can utilize the information on the autoregressive (AR) process of an independent variable to discuss how long its current values persist and therefore how long its effect on the dependent variable lasts. The coefficient of an $AR(p)$ process in an independent variable will describe at which rate its current value diminishes.[3] Therefore, for a bivariate or multiple time series model with the properly prewhitened dependent and independent variables,[4] the coefficient of an $AR(p)$ process in an independent variable can describe at which rate the effect of the independent variable on the dependent variable persists (see e.g. Mackuen, Erikson, and Stimson 1989).

If an independent variable contains an $AR(1)$ process with the coefficient of the lag-1 error term, $\phi$ $(0 < \phi < 1)$, the effect of the independent variable on the dependent variable will persist by the ratio of $\phi$, as lag increases by one.[5] To put it differently, the effect will diminish by the ratio of $1 - \phi$, as lag increases by one. Therefore, the cumulative effect of the independent variable will be $\dfrac{1}{(1 - \phi)}$ times as large as the immediate effect. For example, with $\phi = .3$, the effect of the independent variable will diminish at a rate of 70% per time interval, and the cumulative effect will be about 1.43 times as large as the immediate effect. As $\phi$ approaches one, the effect will persist longer.

An independent variable with no significant $AR(p)$ process has only an immediate effect on the dependent variable. However, when an independent variable contains no significant $AR(p)$ process, its effect on the dependent variable can persist, if the current value of the dependent variable persists for a time interval. Therefore, we can discuss how long the change in the dependent variable that is caused by the independent variable persists by referring to the $AR(p)$ process in the dependent variable.

For an AR(2) process, calculating the persistence rate is complicated (see chapter 2, section 5A). For example, for an independent variable with an AR(2) process whose coefficients of the lag-1 and lag-2 error terms are .3 and .2, respectively, the autocorrelation function (ACF) is .375 at lag 1. Therefore, the effect of the independent variable will diminish by 62.5% after the first lag. Due to the autocorrelation between $\{Y_t\}$ and $\{Y_{t-2}\}$, the ACF at lag 1 is larger than that of an AR(1) process in the above example. In addition, it dies off more slowly than the AR(1) process. The ACFs will be .3125 at lag 2, .175 at lag 3, ..., and .01 at lag 10.[6]

## B. Cross-Correlation

When we assume a causal relationship between two time series, we employ bivariate time series analysis, which is similar to bivariate regression analysis of cross-sectional data. When we do not assume a causal flow from one variable to the other, we calculate cross-correlation, which is similar to the Pearson correlation between two cross-sectional variables. The cross-correlation between two time series at current lag is the Pearson correlation between them. Cross-correlation is different from Pearson correlation in that cross-correlation measures not only how strongly two time series are related with each other at current lag but also how strongly one time series tends to lead or lag the other. The formula for cross-correlation is identical to that for Pearson correlation, except that the former estimates the relationship between two variables that are collected across time through repeated regular observations. The cross-correlation between $\{Y_t\}$ and $\{X_t\}$ when $\{Y_t\}$ leads $\{X_t\}$ by $k$ lags is:

$$r = \frac{\sum_{t=1}^{n-k}\{(Y_t - E(Y_t))\times(X_{t+k}-E(X_t))\}}{\sqrt{\sum_{t=1}^{n-k}(Y_t-E(Y_t))^2}\ \sqrt{\sum_{t=1}^{n-k}(X_{t+k}-E(X_t))^2}}$$

where $t = 1, 2, 3, \ldots, n$; $k = 0, 1, 2, 3, \ldots, (n-1)$; and $E(Y_t)$ and $E(X_t)$ are the means of $\{Y_t\}$ and $\{X_t\}$, respectively.

The cross-correlation between $\{Y_t\}$ and $\{X_t\}$ when $\{Y_t\}$ lags $\{X_t\}$ by $k$ lags is:

$$r = \frac{\sum_{t=1-k}^{n-k}\{(Y_t - E(Y_t))\times(X_{t+k}-E(X_t))\}}{\sqrt{\sum_{t=1-k}^{n-k}(Y_t-E(Y_t))^2}\ \sqrt{\sum_{t=1-k}^{n-k}(X_{t+k}-E(X_t))^2}}$$

where $t = 1, 2, 3, \ldots, n$ and $k = -1, -2, -3, \ldots, -(n-1)$.

As in the above formula, $k$ can be positive or negative. Statistical programs report cross-correlations both for positive $k$ values and for negative $k$ values. When cross-correlation is statistically significant with positive $k$ values, $\{Y_t\}$ is significantly correlated with $\{X_{t+k}\}$, with $\{X_t\}$ shifted $k$ lags forward. In this case, we say that $\{Y_t\}$ leads $\{X_t\}$. When cross-correlation is statistically significant with negative $k$ values, $\{Y_t\}$ is significantly correlated with $\{X_{t-k}\}$, with $\{X_t\}$ shifted $k$ lag(s) backwards. In this case, we say that $\{Y_t\}$ lags $\{X_t\}$. Cross-correlation of a time series with a lagged version of itself is autocorrelation (see chapter 2). The coefficients of the AR($p$) process of a time series are cross-correlations between the time series and lagged versions of itself shifted 1 to $p$ lags backwards.

Even if a cross-correlation procedure is not available, we can calculate Pearson correlations of a time series at current lag with lagged versions of another time series. To determine whether a time series, $\{Y_t\}$, leads or lags another time series, $\{X_t\}$, we first calculate correlations between one time series at current lag, $\{Y_t\}$, and the other time series at lag $k$, $\{X_{t-k}\}$. Then we calculate correlation between the latter at current lag, $\{X_t\}$, and the former at lag $k$, $\{Y_{t-k}\}$. When there are three or more time series, we can repeat this process with every two time series.

The formula for cross-correlation between two time series or for Pearson correlation between two cross-sectional variables does not tell us which of them is the cause. Rather, it describes the direction in which and the extent to which two variables covary. In the social sciences, we are usually unconcerned with identifying which time series is statistically leading and which is lagging. We have our theory about a causal relationship between two time series, instead. We can use cross-correlation to identify lags of a time series that might be useful predictors of another time series in bivariate or multiple time series analysis. For a time series to be a cause (input) of another time series (output), the latter at a certain time should depend on the former up to that time only. Therefore, we usually look at cross-correlations between the dependent variable at current lag and an independent variable at lag $k$ ($k \geq 0$).

We should not determine which time series leads and which lags simply according to the statistical significance of cross-correlations. In the social sciences, the selection of dependent (or lagging) variables and independent (or leading) variables and the selection of lags of independent variables should depend primarily on a theory that we want to test with our model, not on statistical significance of cross-correlations.[7] Cross-correlations reported by statistical programs may be statistically significant even when $k$ is large, for example between a time series at current lag and another time series shifted ten or even twenty years backwards. In addition, cross-correlation may be significant both when $k$ is positive and when $k$ is negative. Cross-correlation may be significant even when the relationship between two time series cannot be theoretically explained. For example, it is not realistic to assume that economic conditions ten years ago influence current presidential popularity or that presidential popularity influences future economic conditions. However, cross-correlations between them may be significant.

## C. Comparability of Values over Time

When we conduct multiple or bivariate time series analysis, we need to consider the possibility that values may not be comparable over time. If a variable does not have fixed minimum and maximum values, its nominal values may be different from its real values. Therefore, when we include time series variables in our model, we often need to adjust their values to make them comparable over time.

For example, the consumer price index needs to be adjusted to correct the problem that prices are not comparable over time due to inflation. We select a base period and adjust prices in terms of the period's general price level. The size of budget also needs to be adjusted by converting it to the dollars of a base year because the raw size of budget may continuously increase simply due to inflation. On the other hand, we do not need to adjust unemployment rate because its minimum and maximum values are fixed and thereby it is comparable over time.

We need somewhat complicated procedures to make a variable comparable over time, if we need to adjust the values of a third variable on which the calculation of the variable is based. Examples are incarceration rates, the number of arrested recidivists per 100,000 inhabitants, and the number of policemen per 100,000 inhabitants. These statistics are calculated based on census results, and therefore they are likely to be increasingly overstated or understated as the census period nears the end.

For example, the incarceration rate is calculated by dividing the number of state and federal prisoners by the total number of resident population. Since census data are identical for every ten-year period, the total population is likely to be increasingly understated as the census period nears the end, if there is a upward trend in the population size. Consequently, the incarceration rate is likely to be increasingly overstated. If there is a downward trend in the population size, the incarceration rate is likely to be increasingly understated. Thus, we need to adjust the population for every year between every two consecutive censuses. Since we cannot know for sure the total population for every year, we increase (or decrease) it each year by one-tenth of the difference between every two consecutive censuses. For example, the adjusted total population of 1983 will be equal to the total population in the 1980 census plus three-tenths of the difference between the 1980 census and the 1990 census. For 1989, it will be equal to the total population in the 1980 census plus nine-tenths of the difference between the 1980 census and the 1990 census. For 1991, it will be equal to the total population in the 1990 census plus one-tenth of the difference between the 1990 census and the 2000 census.

### D. The Case of Monthly Violent Crime Rates, 1983–1992

In chapter 2, we analyzed the time-dependent behavior of the monthly violent crime rates from 1983 to 1992. Many factors will influence crime rates. Here, we explore the relationship between two economic conditions (unemployment rate and inflation) and monthly violent crime rates. These two factors have been frequently employed to measure economic adversity. Individuals are expected to become more likely to commit crimes when they experience economic adversity. Therefore, as the unemployment rate increases and inflation becomes worse, the violent crime rate is expected to increase. Inflation was measured with consumer price index adjusted with the period from 1982 to 1984 as the base period to correct the problem that prices are not comparable over time due to inflation.

First, we check the simple relationship between each of these economic conditions and monthly violent crime rates without assuming a causal flow from economic conditions to violent crime rates. As mentioned above, one time series may be significantly related to the other not only at current lag but also at lag $k$ ($k > 0$). In the social sciences, we do not check all these possible relationships. We check only theoretically plausible relationships.

Regarding the monthly violent crime rates and the two economic conditions, we may find a statistically significant relationship between current crime rates and two economic conditions in previous years. However, an estimated relationship between economic conditions in previous years and crime rates is likely to be conditioned by current economic conditions. For example, those whose economic conditions were not good in the previous year but improved in the current year are less likely to commit crimes, compared with those whose economic conditions were good in the previous year but not good in the current year. Those whose economic conditions were good in the previous year but worsened a lot in the current year are more likely to commit crimes, compared with those whose economic conditions were not good in the previous year but good in the current year. We may calculate cross-correlation between current crime rates and the difference between current economic conditions and previous economic conditions, which measures whether economic conditions improved or worsened. However, worsened economic conditions may be still good, while improved economic conditions may be still bad.

Here, we assume that economic conditions and violent crime rates tend to go hand in hand, rather than that the former lead the latter. Therefore, we focus on the economic conditions in the current year and estimate their relationship with the monthly violent crime rates. The estimated relationship may not be highly accurate because the relationship may be conditioned by the economic conditions in the previous year. However, the estimated relationship can tell us whether and how strongly violent crime rates respond instantaneously to economic conditions.

Cross-correlations were calculated for all 120 lags. Table 8 reports the estimated cross-correlations between current crime rates and each of the two economic conditions shifted twelve lags forward or backward. As shown in table 8, inflation is most strongly and positively correlated with violent crime rates at current lag, tapering in both directions from that peak. This indicates that violent crime rates tend to instantaneously respond to inflation, increasing as inflation gets more serious. Since cross-correlations are significant at negative as well as positive lags, both inflation and crime rates lag and lead each other.

Most cross-correlations (after the fifth lag) between crime rates and unemployment rates shifted forward are insignificant, indicating that unemployment rates tend to lead crime rates. However, the cross-correlation between crime rates and unemployment rates is not in accord with the general expectation. There are maximum negative values at −24 (lag 24), tapering in both directions from that peak, indicating that an above- (below-) average unemployment rate is likely to lead to a below- (above-) average crime rate about 24 months later.

A possible explanation for this unexpected relationship is offered by the scatter plot of the unemployment rates from 1983 to 1992 (figure 39). According to figure 39, unemployment rates showed a quadratic trend,

TABLE 8. Cross-Correlation of Monthly Violent Crime
Rates, 1983–1992, with Monthly Economic Adversity

| Lag | Unemployment rate | Inflation |
|---|---|---|
| –12 | –.579 (.096) | .638 (.096) |
| –11 | –.57 (.096) | .663 (.098) |
| –10 | –.56 (.095) | .686 (.095) |
| –9 | –.54 (.095) | .703 (.095) |
| –8 | –.518 (.094) | .718 (.094) |
| –7 | –.484 (.094) | .728 (.094) |
| –6 | –.455 (.094) | .74 (.094) |
| –5 | –.432 (.093) | .758 (.093) |
| –4 | –.414 (.093) | .778 (.093) |
| –3 | –.401 (.092) | .801 (.092) |
| –2 | –.389 (.092) | .822 (.092) |
| –1 | –.388 (.092) | .848 (.092) |
| 0 | –.371 (.091) | .865 (.091) |
| 1 | –.334 (.092) | .855 (.092) |
| 2 | –.295 (.092) | .846 (.092) |
| 3 | –.262 (.092) | .825 (.092) |
| 4 | –.228 (.093) | .804 (.093) |
| 5 | –.194 (.093) | .778 (.093) |
| 6 | –.161 (.094) | .75 (.094) |
| 7 | –.13 (.094) | .728 (.094) |
| 8 | –.102 (.094) | .705 (.094) |
| 9 | –.074 (.095) | .687 (.095) |
| 10 | –.05 (.095) | .673 (.095) |
| 11 | –.027 (.096) | .664 (.096) |
| 12 | –.01 (.096) | .65 (.096) |

NOTE. Numbers in parentheses are standard errors.

$10.224 - 0.13\text{Time} + 0.001\text{Time}^2$, during this period. Unemployment rates
fell exponentially during about the first six years and then rose exponentially.
Cross-correlation, which is Pearson correlation between two time series, is
only appropriate as a measure of linear relationship. Therefore, cross-correlations between crime rates and unemployment rates actually measured relationships between the monthly violent crime rates with a positive linear
trend (see chapter 2) and the unemployment rates with a weak negative linear
trend, $7.965 - 0.019\text{Time}$. Consequently, above-average unemployment rates
tended to be compared with below-average crime rates, while below-average
unemployment rates tended to be compared with above-average crime rates.

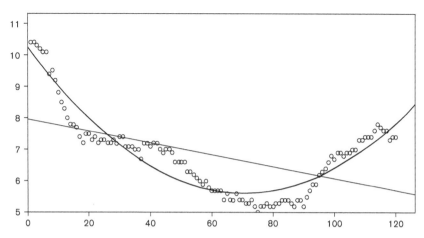

FIGURE 39. Monthly unemployment rates, 1983–1992.

The unexpected relationship may be caused by the Pearson correlation coefficient's characteristic that it is concerned with the distance of two variables from their means (Erdem, Ceyhan, and Varli 2014). If both variables are above (or below) their means, the sum of multiplication of the two variables' deviations from their means will positively contribute to the numerator in the correlation coefficient, even when the variables move in opposite directions. On the other hand, if one variable is above its mean and the other variable is below its mean, the sum of multiplication of the two variables' deviations from their means will negatively contribute to the numerator and hence to the correlation coefficient, even when the variables move in the same direction. The latter is likely to be the case according to figure 39. The first and last quarters of unemployment rates are likely to be above average, while the second and third quarters are likely to be below average. On the other hand, the first half of crime rates are likely to be below average, while the second half are likely to be above average. Thus, the first and third quarters of unemployment rates and corresponding crime rates are likely to contribute negatively to the correlation coefficient.

To test whether higher unemployment rates and inflation increase violent crime rates, we need to conduct multiple time series analysis. The dependent variable is monthly violent crime rate, and the independent variables are monthly unemployment rate and monthly consumer price index. For the reasons mentioned above, we used both independent variables at current lag.

Table 9 reports the result of the multiple time series analysis. According to table 9, as inflation became more serious, the violent crime rates tended to

TABLE 9. Monthly Economic Adversity and Monthly Violent Crime
Rates, 1983–1992

|  | Coefficient | Standard error |
|---|---|---|
| Unemployment rate | .059 | .318 |
| Consumer price index | .505*** | .03 |
| Constant | −6.833 | 4.917 |
| $N$ | 120 | |
| Adj. $R^2$ | .744 | |

***$p < .001$ (one-tailed test).

increase during the ten-year period from 1983 to 1992. A two-point increase in consumer price index tended to cause about a one-point increase in violent crime rates. Unemployment rate was not significantly related to the violent crime rates.

Visual inspection of the dependent variable and the two independent variables supports this result. According to figures 39 and 40, monthly unemployment rates and monthly consumer price index contain a quadratic trend and a linear trend, respectively. Thus, the change of unemployment rate does not seem to match well with the change of violent crime rate, which shows a linear upward trend (see figure 9, chapter 2, page 20).

The residuals of our model (table 9) contain a significant autocorrelation. For example, the observed Durbin-Watson $d$ test statistic is 0.736. Since the number of cases is 120 and the number of independent variables including the constant is 3, the lower bound of the Durbin-Watson $d$ statistic is 1.613 when the level of significance is .05 (see table A1, appendix 4, page 205). This lower bound is higher than the observed $d$, indicating that residuals contain a significant positive autocorrelation.

Visual inspection of the residuals supports this interpretation. According to figure 41, the ACFs show an abrupt, roughly geometric decay, while the PACFs spike several times, suggesting the presence of an AR process in the residuals. If the residuals are a realization of an independent and identically distributed (IID) sequence, we expect that less than about 5% of the sample PACFs at lag 1 or higher will fall outside the 95% confidence bounds. Since we have 120 observations, we expect that less than 6 sample PACFs will fall outside the bounds, ±0.179. In figure 41, the ACFs show a cyclical pattern, with each cycle lasting for twelve lags. Six PACFs fall outside the bounds. The first PACF most clearly falls outside the 95% confidence bounds. The PACFs

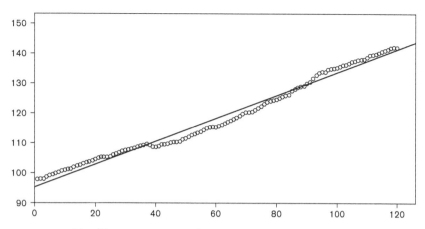

FIGURE 40. Monthly consumer price index, 1983–1992.

at lags 12 and 13 also clearly fall outside the bounds, and then PACFs cut off after lag 13. A bit of action comes at lags 3, 4, and 6. This is a typical behavior of a time series with seasonality of lag 12. We had better not count these PACFs to determine the order of an AR process. The PACFs that fall outside the confidence bounds, except the first one, may be produced by seasonal components. With the deseasonalized dependent variable, the ACFs no longer show a cyclical pattern, and only the first PACF clearly spikes (figure 42). In sum, according to the ACF and PACF correlograms, the residuals appear to contain an AR(1) process with seasonality of lag 12.

Thus, the multiple time series model in table 9 violates an important assumption for the least square estimates to be robust: no autocorrelation in residuals. We can deal with this violation in different ways. We may whiten the dependent variable and independent variables before we analyze their relationship (see section 2). We may respecify our model by including additional independent variables, by changing the functional form of the relationship between the dependent variable and an independent variable, or by incorporating the AR process of the residuals into our model (see section 4).

(2) PREWHITENING

A time series is collected across time through repeated regular observations on a single unit of analysis. Therefore, either or both variables in cross-correlation analysis or in bivariate or multiple time series analysis may contain

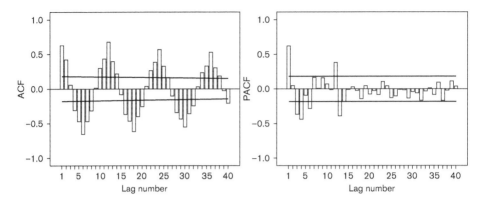

FIGURE 41. ACF and PACF of residuals of the multiple time series analysis of the unwhitened data.

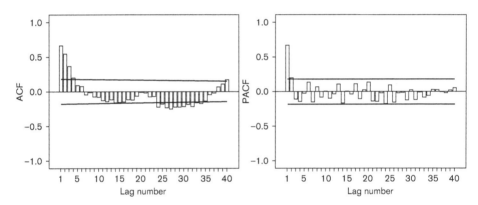

FIGURE 42. ACF and PACF of residuals of the multiple time series analysis of the deseasonalized unwhitened data.

systematic patterns, AR($p$) and/or MA($q$) processes (see chapter 2). These systematic patterns may cause autocorrelation in residuals of a bivariate or multiple time series model.

In addition, systematic patterns may bias cross-correlation coefficients and the OLS regression coefficients in bivariate or multiple time series analysis. AR($p$) and/or MA($q$) processes in one of the two variables may change the degree to which the variable covaries with the other variable. When both variables contain systematic patterns, the concordance or discordance of these systematic patterns can disguise the actual relationship between the two variables. The concordance of these systematic patterns will produce an overstated coefficient estimate, while the discordance of these systematic patterns will produce an understated coefficient estimate. The degree of bias of an estimated relationship will depend on the degree to which their systematic patterns are concordant or discordant with each other. In addition, the standard error of the estimated coefficient and the coefficient of determination are likely to be lower or higher than they would be without these systematic patterns of residuals.

For example, a president's popularity may increase or remain high for several months, and then it may decrease for the next several months. Economic conditions, such as GNP and unemployment rate, may also continue to improve for several months and then worsen for the next several months. The concordance of this increase and decrease between presidential popularity and economic conditions will be reflected in an estimated positive relationship between them. Consequently, we cannot confidently conclude that presidential popularity improves as economic conditions improve.

An omitted third factor may also produce a biased coefficient estimate of the independent variable in multiple time series analysis, as in multiple OLS analysis of cross-sectional data.[8] Because an omitted third factor is related to model specification, we control this third factor by including it in our model. Different from the intervening role of a third factor, the concordance or discordance of the systematic patterns of the dependent and an independent variable may not be theoretically meaningful. Therefore, we need to control out these systematic patterns.

The process of controlling out these systematic patterns is called *prewhitening* because we make the variables' disturbances consist of white noise before we analyze their relationship. We can prewhiten a time series with the estimate-and-then-eliminate method (see chapter 2). We deseasonalize and/or detrend a time series variable and then estimate a systematic temporal pattern of the deseasonalized and/or detrended series. Then we eliminate the estimated systematic temporal pattern from the original time series variable.

Alternatively, we can smooth the dependent variable and an independent variable to improve the accuracy of the estimated coefficient of the independent variable (see chapter 5). If we do not need to estimate systematic patterns of the two variables, smoothing the two variables will be an efficient alternative strategy to control the temporal coincidence of the two variables' systematic patterns. With smoothing, however, we may not be sure about whether we accurately eliminate only systematic patterns from each variable.

## A. Two Approaches

In bivariate or multiple time series analysis, both the dependent variable and an independent variable can contain systematic patterns. In this case, we can take two different approaches to prewhiten the two variables. One is the *systems* approach (Chatfield 2004). We prewhiten an independent variable with the estimated systematic pattern, $\text{ARMA}(p, q)$, of its residuals, and then we filter the dependent variable with the same $\text{ARMA}(p, q)$ process. The other is the *equal footing* approach (Brockwell and Davis 2002; Chatfield 2004). We prewhiten the dependent variable and an independent variable individually with an estimated systematic pattern of each variable's residuals. In cross-correlation analysis, we prewhiten variables in the same way.

When applying the systems approach, we prewhiten the lead (input) variable and filter the lag (output) variable. In the systems approach, a causal flow from an independent variable to the dependent variable is emphasized. By eliminating the estimated systematic pattern from an independent variable, we remove the possibility for the dependent variable to be explained by its temporal concordance or discordance with the independent variable's systematic pattern. We eliminate systematic patterns from the independent variable but not from the dependent variable. Instead, by filtering the dependent variable with the estimated systematic pattern of the independent variable, we eliminate only the portion of the dependent variable's variation that can be explained by the estimated-and-then-eliminated systematic pattern of the independent variable. We leave the other portion untouched. If the dependent variable has the same systematic pattern, it will also be whitened by the filtering. If not, the dependent variable may still contain systematic patterns after the filtering procedure.

Although both approaches have been employed, the equal footing approach may be preferred to the systems approach for several reasons.[9] First, when we filter the dependent (or output) variable with an estimated systematic pattern of an independent (or input) variable, we may not completely eliminate the systematic patterns of the dependent variable. The remaining systematic patterns of the dependent variable may have significant concordance or discordance with the regular time-dependent components, such as trend, of the independent variable. If so, the estimated relationship between the two variables may be biased even after we prewhiten the independent variable and filter the dependent variable. Second, we cannot determine a prewhitening filter for the dependent variable when we have two or more independent variables that contain different systematic patterns. Third, we will spend almost identical time for both approaches. For the systems approach, we may save time because we do not estimate the systematic pattern of the dependent variable. However, the amount of time saved will usually be small because we filter the dependent variable with the estimated systematic pattern of an independent variable.

In sum, when we prewhiten a time series variable, we transform its residuals into white noise. This way, we can eliminate the possibility that two variables are observed to be positively or negatively related to each other simply because of the concordance or discordance of their residuals' systematic patterns. However, when we prewhiten variables in bivariate or multiple time series analysis, we should keep in mind that prewhitening eliminates systematic temporal patterns from the dependent variable and/or independent variables, not necessarily the autocorrelation of the residuals of our time series model. Prewhitening often solves the autocorrelated-residuals problem. However, the autocorrelation of residuals in our time series model can be caused by various factors, and it may remain after we have prewhitened the dependent variable and independent variables. Therefore, we need to check that no serious autocorrelation exists in residuals at the end of time series analysis (see section 4, below).

## B. Trend and Seasonality

When we prewhiten a time series in the social sciences, we usually deal with residuals, the disturbances that are left over after we eliminate seasonality and/or a trend of the original time series, not with the original series. Trends and seasonality are regular temporal patterns, which we may need to consider to accurately explain the relationship between the dependent variable and an independent variable. Trends and seasonality are different from systematic patterns of a deseasonalized and detrended time series. The latter are also systematic temporal patterns of a variable, but they usually do not convey any theoretical meanings. They simply represent nonrandom fluctuations.

Whether eliminating seasonality and trends from the dependent variable and an independent variable helps us estimate their relationship accurately may be controversial. When the prewhitening procedure includes the deseasonalization and detrending procedures, prewhitening may virtually eliminate most of the variation of the dependent variable that we expect our model to explain and the variation of the independent variables that we expect to explain the dependent variable. We may not conclude with confidence that the insignificant coefficient of the independent variable evidences the lack of relationship between the two variables.

In the social sciences, the decision should be made theoretically according to whether we should explain the trend and seasonality of the dependent variable. If trends and seasonality do not convey any theoretical meanings with regard to our research question, we should eliminate trends and seasonality from variables for more accurate model estimation. In the natural sciences, trends generally do not convey theoretical meanings. For example, we simply observe the increase or decrease of volcanic activities. Such increase or decrease may be just an irregularly repeated pattern that is not caused by independent variables in our model. In this case, we may need to control out trends to estimate our model more accurately. Otherwise, estimated coefficients may be biased, especially if independent variables are also trending.

In the social sciences, eliminating trends may be more controversial than eliminating seasonality. A trend of the dependent variable is generally an important part of the dependent variable's variation that we should explain with our model. For example, we attempt to explain why crime rates have increased or decreased in multiple time series analysis. In interrupted time series analysis (see chapter 7), we test whether crime rates decreased after we implemented a tougher crime-control policy. By removing increasing or decreasing trends from the dependent variable and/or an independent variable, we are likely to flatten the dependent variable's variation that we should explain with the independent variable and the independent variable's variation that can explain the dependent variable. Consequently, estimated coefficients may not accurately tell us what we want to learn from them.

If we determine that trends of the dependent variable and independent variable(s) are not theoretically important in testing our research questions, controlling out trends from our analysis is easy. In bivariate or multiple time series analysis, we can control out trends by detrending the dependent variable and independent variables individually (see chapter 2).

We can also take two different approaches with regard to how to deal with seasonal components of variables in our model. First, we may consider the seasonality of the dependent variable as an important part of the dependent variable that we should explain with our model. The dependent variable's seasonal fluctuations may be caused at least partly by independent variables. In addition, we may consider the seasonality of an independent variable as an important part of the variable's variation, with which we should explain the dependent variable's variation. In this case, after eliminating seasonal components, we may inaccurately estimate the relationship between the dependent variable and independent variables.

For example, crimes tend to be committed more frequently in warm periods, especially around summer, and less frequently during cold periods (see table 1, chapter 2, page 26). These seasonal fluctuations can be parts of the dependent variable's variation that can be influenced by economic conditions. Seasonal fluctuations may become less clear as economic adversity becomes more serious because potential criminals may become more likely to commit crimes, regardless of weather conditions. In this case we may need to leave out seasonal components and estimate the effect of independent variables on the variation of the dependent variable with seasonality.

Second, we may consider the seasonality of the dependent variable and/or of an independent variable as external sources of variation that could complicate the identification process (Makridakis, Wheelwright, and McGee 1983, 485–501). The seasonality of the dependent variable may simply represent temporal regularities that we do not expect our independent variables to explain. For example, seasonal fluctuations of monthly violent crime rate may be simply repeated every year, whether economic conditions are improved or worsened. Therefore, if crime rates are not deseasonalized, we leave out the variation of the dependent variable that we know can be represented simply by seasonal fluctuations and that we do not expect economic adversity, such as unemployment and inflation, to explain. Seasonal components increase and decrease periodically, and their influence on the estimated long-term relationship between the dependent variable and an independent variable is likely to be cancelled. Therefore, the estimated coefficient of the independent variable may not be biased. However, seasonal components are likely to bias the standard error of the estimated coefficient and the coefficient of determination of the model.

When both the dependent variable and an independent variable contain seasonal components, the concordance or discordance between their seasonal components may bias the estimated relationship between the two variables. For example, economic adversity may tend to be most (or least) serious from July to September and least (or most) serious from December to February during the period of study. We usually cannot generalize this type of seasonal fluctuations of economic conditions. However, the concordance (or discordance) between economic conditions and violent crime rates in terms of these seasonal fluctuations is likely to produce an overestimated (or underestimated) positive relationship or underestimated (or overestimated) negative relationship between the two time series.

We should apply either of these two approaches both to the dependent variable and to an independent variable, when both variables contain seasonal components. If we keep the dependent variable's seasonality, we also need to keep the independent variable's seasonality. If we deseasonalize the dependent variable, we also need to eliminate seasonal components from independent variable(s). Otherwise, estimated relationships between the dependent variable and an independent variable may be biased.

If an independent variable, but not the dependent variable, contains seasonal components, we usually had better eliminate seasonal components from the independent variable, even when the seasonality itself is theoretically meaningful. The dependent variable's variation does not contain a portion that can be explained by the seasonal components of the independent variable.

If the dependent variable, but not an independent variable, contains seasonal components, we can deal with the seasonality in two different ways. First, we can deseasonalize the dependent variable and then estimate the model. When we forecast future values of the dependent variable, we adjust forecasted values with the estimated seasonal components (see chapter 4). Second, we can directly model the seasonality of the dependent variable with a dummy variable scheme. For example, when the dependent variable consists of quarterly or monthly seasonal components, we can include three or eleven indicator variables, respectively. If there are quarterly seasonal components, we include $Q_1$, $Q_2$, and $Q_3$. $Q_1$ will be 1 for the first quarter and 0 for the other quarters. $Q_2$ will be 1 for the second quarter and 0 for the other quarters. $Q_3$ will be 1 for the third quarter and 0 for the other quarters. For the monthly seasonal indicators, we include $M_1$ to $M_{11}$. $M_1$ will be 1 for January and 0 for the other months, $M_2$ will be 1 for February and 0 for the other months, and so on. We should not include all four quarters ($Q_1$ to $Q_4$) or all twelve months (January to December) in the dummy variable scheme to avoid the perfect multicollinearity (zero tolerance) problem.

The dummy variables will control variations in the mean values among the quarters or the months. Their coefficients will denote differences in the mean values between the three quarters or the eleven months included in the model and the excluded category (the fourth quarter or December in the above examples) in the dummy variable scheme. Therefore, the coefficients of the dummy variables will be similar to the seasonal components identified with the additive method. Both of them describe how much each seasonal component pulls up or down the overall level of a time series on average.

The estimated relationship between the dependent variable and independent variables will be similar whether we deseasonalize the dependent variable or include a dummy variable scheme to control seasonal variations in our model. The procedure to predict values will be virtually identical: we predict values with the estimated coefficients of independent variables, and then we adjust the predicted values according to the estimated seasonal components or according to the estimated coefficients of the seasonal (quarter or month) indicators.

With a dummy variable scheme, the number of independent variables will increase by three (for quarterly data) or by eleven (for monthly data). This may not be desirable, especially if our time series analysis includes a small number of cases. In addition, dummy variables may cause serious multicollinearity problems.

### C. The Case of Monthly Violent Crime Rates, 1983–1992

Before we conclude that violent crime rates tend to increase as inflation becomes worse, we need to suspect the possibility that the estimated relationships between the dependent variable and each of the two independent variables may be biased by their systematic patterns. We already identified an AR(2) process as the systematic pattern of the deseasonalized and detrended monthly violent crime rates (see chapter 2). Unemployment rate and consumer price index are also likely to contain systematically patterned residuals. In this case, the observed significant or insignificant relationship between the dependent variable and each of the two independent variables may be caused by the concordance or discordance of the systematic patterns.

To check whether unemployment rate and consumer price index contain systematically patterned residuals, we estimated and eliminated trends from monthly unemployment rate and monthly consumer price index, respectively. According to figures 43 and 44, the two detrended variables still show nonrandom fluctuations, with long stretches of plots on the same side above or below the trend line that equals zero across time. These long stretches indicate that the two detrended variables are serially correlated. Because the dependent variable and the two independent variables contain autocorrelations, the estimated relationships in table 9 are likely to be biased. To estimate the causal relationship between the two independent variables and the

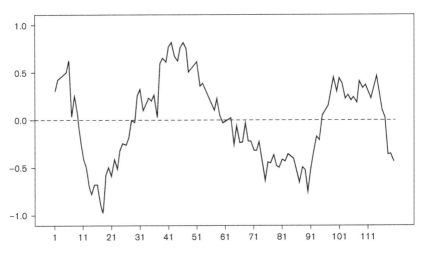

FIGURE 43. Detrended monthly unemployment rates, 1983–1992.

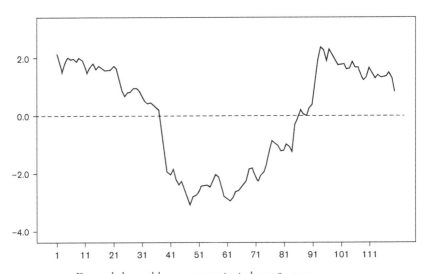

FIGURE 44. Detrended monthly consumer price index, 1983–1992.

dependent variable more accurately, we need to prewhiten all three variables.

According to the ACF and PACF correlograms of the detrended monthly unemployment rates (figure 45), the ACFs rapidly decay, while the PACF at lag 1 crosses the 95% confidence bounds. Therefore, the systematic pattern of the detrended monthly unemployment rate can be represented by an AR(1)

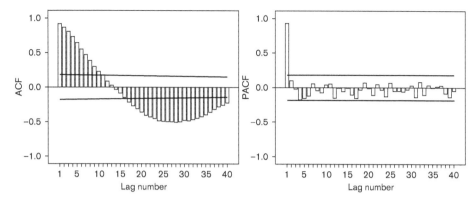

FIGURE 45. ACF and PACF of detrended monthly unemployment rates, 1983–1992.

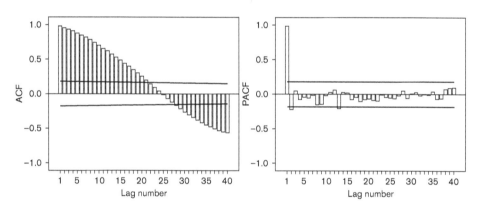

FIGURE 46. ACF and PACF of detrended monthly consumer price index, 1983–1992.

process.[10] The estimated coefficient is 0.992 (standard error 0.008). According to the ACF and PACF correlograms of the detrended consumer price index (figure 46), the ACFs rapidly decay, while the PACFs at lag 1 cross the 95% confidence bounds. The second PACF bar crosses the confidence bound only slightly. Therefore, the systematic pattern of the detrended consumer price index may be represented by an AR(1) process. The estimated AR coefficient is 0.987 (0.012).[11]

After we removed these systematic patterns from the two detrended independent variables, the two variables are whitened (for the whitened dependent variable, see chapter 2). Their residuals do not show any systematic temporal patterns (figures 47 and 48). In addition, less than three ACFs and

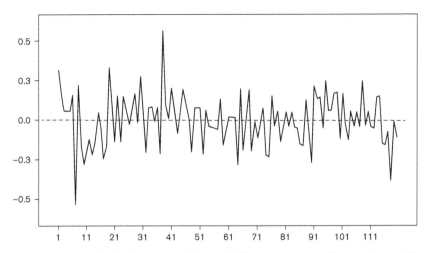

FIGURE 47. Residuals obtained by fitting AR(1) process to the detrended monthly unemployment rates, 1983–1992.

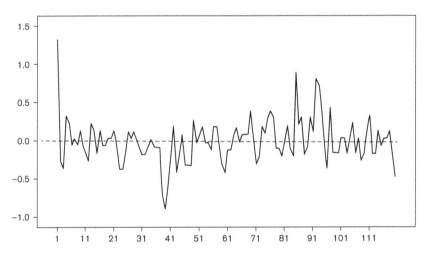

FIGURE 48. Residuals obtained by fitting AR(1) process to the detrended monthly consumer price index, 1983–1992.

PACFs cross the 95% confidence bounds (figures 49 and 50). We prewhiten these three variables by eliminating each variable's systematic residual pattern. By prewhitening the dependent variable and the two independent variables, we make the estimated coefficients of the two independent variables no longer be an artifact of the systematic residual patterns of these three variables.

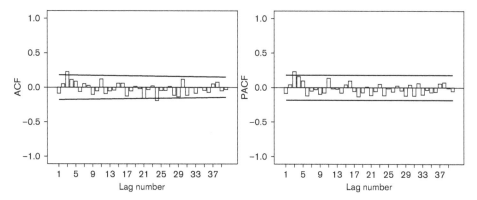

FIGURE 49. ACF and PACF of residuals obtained by fitting AR(1) process to the detrended monthly unemployment rates, 1983–1992.

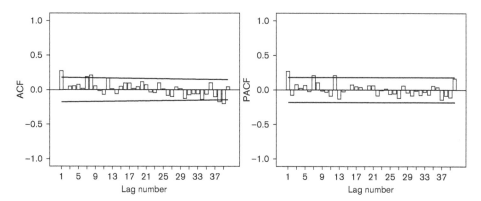

FIGURE 50. ACF and PACF of residuals obtained by fitting AR(1) process to the detrended monthly consumer price index, 1983–1992.

Table 10 reports the result of the multiple time series analysis of the prewhitened data. Like the model in table 9, the first column includes the dependent variable and the two independent variables, but these variables were prewhitened. The dependent variable was deseasonalized because only the dependent variable contains seasonal components. In the second column, the dependent variable was prewhitened but not deseasonalized. The second column additionally includes the eleven month indicators, instead. In the third column, the dependent variable and the two independent variables were not prewhitened. The dependent variable was not deseasonalized. That is, this column adds the eleven month indicators to the model in table 9. Since the month indicators do not cause serious multicollinearity problems

TABLE 10. Prewhitened Monthly Economic Adversity and Monthly Violent Crime Rates, 1983–1992

| | Column 1 | Column 2 | Column 3 |
|---|---|---|---|
| Unemployment rate | .024 (.13) | .024 (.137) | .068 (.166) |
| Consumer price index | .492*** (.012) | .492** (.012) | .495*** (.016) |
| January | | −1.292 (.665) | −.956 (.922) |
| February | | −5.859*** (.665) | −5.235*** (.922) |
| March | | −1.308 (.665) | −.673 (.922) |
| April | | −1.801** (.665) | −1.293 (.921) |
| May | | 1.714* (.665) | 2.102* (.921) |
| June | | 2.58*** (.665) | 2.832** (.921) |
| July | | 6.551*** (.665) | 6.654*** (.921) |
| August | | 6.669*** (.664) | 6.867*** (.921) |
| September | | 2.99*** (.664) | 3.087*** (.920) |
| October | | 3.002*** (.664) | 2.957*** (.920) |
| November | | −1.589* (.664) | −1.598* (.920) |
| Constant | −5.112** (2.039) | −6.116** (2.184) | −6.958** (2.651) |
| N | 120 | 120 | 120 |
| Adj. $R^2$ | .955 | .963 | .930 |

*$p < .05$, **$p < .01$, ***$p < .001$ (one-tailed test).

NOTE. Numbers in parentheses are standard errors.

in columns 2 and 3, with the tolerance levels greater than .5, we can confidently interpret the results.

According to the first column, after the serial correlations were eliminated from the dependent variable and from the two independent variables, the result remains almost identical to that in table 9. Consumer price index is statistically significant and positively related with the violent crime rate. The effect of a ten-point increase in inflation is about a five-point increase in the violent crime rate. Violent crime rates did not tend to respond instantaneously to unemployment rate.

According to the third column, after we controlled out the monthly variations of violent crime rate without prewhitening the dependent variable and the two independent variables, inflation's coefficient became closer to the coefficient in columns 1 and 2, compared with that in table 9. Inflation's coefficient reduced further, after we prewhitened the deseasonalized dependent variable and the two independent variables (column 1) or after we prewhitened the three variables and controlled out seasonality with the month indicators (column 2). The seasonal components in the dependent variable and the systematic residual patterns in the three variables appear to inflate the coefficient of inflation and its standard error slightly.

According to the first and second columns, the coefficients of the two independent variables are identical whether we deseasonalized the dependent variable or included the eleven month indicators. The estimated coefficients of the month indicators are similar to the seasonal components estimated with the additive method (see table 1, chapter 2, page 26). They are slightly closer to the estimated seasonal components in table 1 when we prewhitened the dependent and independent variables (column 2) than when we did not (column 3).

Thus, systematic residual patterns in the three variables and the seasonal components of the dependent variable do not appear to significantly distort the estimated coefficients of the multiple time series model. However, compared with table 9, the standard errors of the estimated coefficients decreased when we controlled out the seasonal components of the dependent variable (column 3) and prewhitened variables (columns 1 and 2). In addition, according to the adjusted $R^2$, the amount of the dependent variable's variation that was explained by the two independent variables increased when we controlled out the seasonal components of the dependent variable (column 3) and prewhitened variables (columns 1 and 2).

Including the month indicators or deseasonalizing the dependent variable did not help us solve the autocorrelated-residuals problem. The residuals for column 3 still contain a significant positive autocorrelation, with the Durbin-Watson $d$ equal to 0.595. Once we prewhitened the dependent variable and the two independent variables, the residuals of the model (the first and second columns in table 10) do not show a significant autocorrelation. The observed Durbin-Watson $d$ test statistic is 1.862 for column 1. It is 1.883 for column 2. They are higher than the lower bounds at $\alpha = .05$, 1.613 when $k = 3$ and 1.371 when $k = 14$ (see table A1, appendix 4, page 205). Therefore, we do not reject the null hypothesis that the residuals are a realization of an IID sequence.

Visual inspection of the residuals supports these interpretations. According to figure 51, when the serial correlations are removed from the dependent variable and from the two independent variables, we cannot find evidence for the AR process in the residuals of the multiple time series analysis (column 1). Less than 5% of the 120 observations (two ACFs and no PACFs) fall slightly outside the confidence bounds. As in this case, prewhitening may help us not only to estimate the relationship between the dependent variable and an independent

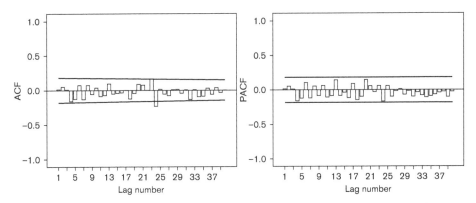

FIGURE 51. ACF and PACF of residuals of the multiple time series analysis of the prewhitened time series data.

variable more accurately but also to solve the autocorrelated-residuals problem.

If we detrend the whitened and deseasonalized dependent variable and the whitened independent variables, both independent variables' coefficients are statistically insignificant. In addition, the model does not explain the dependent variable's variation at all, as its adjusted $R^2$ is zero, and the equation's level of significance is .86. This is an expected result, as we virtually eliminated most of the variation of the dependent variable that we expect our model to explain and the variation of the independent variables that we expect to explain the dependent variable.

As mentioned above, we do not need to check multivariate relationships among monthly violent crime rates, monthly unemployment rates, and monthly consumer price index. We cannot theoretically justify causal flows from monthly violent crime rate to unemployment rate or consumer price index. Even if we find a statistically significant relationship between violent crime rates at lag 1 and unemployment rates or consumer price index at current lag, it may be appropriate to conclude that violent crime rate simply tended to lead, rather than cause, unemployment rate or consumer price index during the observation period. If we can theoretically justify causal flows from the dependent variable to the two independent variables, we can repeat the above estimation process to estimate bivariate time series models with each of the two independent variables as the dependent variable and the monthly violent crime rate as the independent variable. We may build a path model with each of the three variables at current lag as the dependent variable and the others at lag $k$ ($k > 0$) as the independent variables.

## (3) MULTIPLE TIME SERIES ANALYSIS WITH LAGGED VARIABLES

### A. Granger Causality

In bivariate or multiple time series analysis, observations are repeated on a single unit of analysis. Therefore, we need to discern changes of the dependent variable caused by independent variables from the simple continuity of the dependent variable. A related concept is the *Granger causality* (Granger 1969). A time series is considered a Granger cause of another time series if the former predicts the latter significantly better than the latter predicts itself. In other words, a time series, $\{X_t\}$, needs to significantly explain another time series, $\{Y_t\}$, after the continuity of $\{Y_t\}$ is controlled out. To test this point, we include a lagged dependent variable on the right-hand side:[12]

$$(8) \qquad Y_t = b_0 + b_1 Y_{t-1} + b_2 X_{t-1} + E_t$$

Several points deserve our attention with regard to the lagged dependent variable in equation (8). First, the coefficient of the lagged dependent variable, $b_1$, does not have a causal interpretation, although the variable is on the right-hand side. It simply represents the continuity of the dependent variable between two adjacent time points.

Second, the effect of excluded important independent variables, if any, on the dependent variable will be reflected in $b_1$. Therefore, we can additionally control unmeasured shocks with the lagged dependent variable. In addition, the lagged dependent variable(s) will help us eliminate autocorrelation in the residuals, if the autocorrelation is caused by excluded important independent variables (see section 4, below).

Third, $b_1$ may not be identical to the AR(1) coefficient, when $\{X_{t-1}\}$ is a Granger cause. The concept of the Granger causality was originally developed for a stationary time series with a constant autocorrelation structure but without a trend and seasonality (Granger 1969). In this case, $b_1$ may be the coefficient of an AR(1) process if $\{X_{t-1}\}$ is not a Granger cause and thereby does not significantly contribute to equation (8). However, if $\{X_{t-1}\}$ is a Granger cause, equation (8) is a multiple OLS regression model and $b_1$ and $b_2$ are simultaneously estimated. In this model, terms corresponding to the variance of both $\{Y_{t-1}\}$ and $\{X_{t-1}\}$ and a term corresponding to the covariance of $\{Y_{t-1}\}$ and $\{X_{t-1}\}$ appear in the formula for their slopes. Therefore, the coefficient of $\{Y_{t-1}\}$, $b_1$, represents the relationship between $\{Y_t\}$ and $\{Y_{t-1}\}$, when $\{X_{t-1}\}$ is being held constant.

In addition, $b_1$ may not be identical to an AR(1) coefficient, even when $\{X_{t-1}\}$ is not a Granger cause. The continuity of a variable is caused not only by its theoretically unexplainable AR process but also by other various factors. For example, party identification tends to be highly stable in the U.S. (see e.g. Abramson and Ostrom 1991; Bartels et al. 2010; Campbell et al. 1960; Green et al. 2001). When we prewhitened quarterly macro-level party identification collected by Gallup from 1953 to 1996 by eliminating its strong AR(1) process ($\phi = .9$), the coefficient of a bivariate OLS regression between party identification and its lagged term was reduced from 0.92 to 0.41, but it was highly significant ($p = .000$, one-tailed). When we removed its linear trend ($-0.035, p = .000$) as well as its AR(1) process, the coefficient of a bivariate OLS regression fell to 0.15, but it was still significant ($p = .027$). This indicates that the continuity of party identification is not explained simply by its AR process. For example, other factors such as strong affective attitudes toward the two major political parties may contribute to the continuity. In this case, the continuity of the dependent variable that we control in the Granger causality model may not be just an AR process, and therefore $b_1$ is not an AR coefficient. Instead, it represents the relationship between $\{Y_t\}$ and $\{Y_{t-1}\}$ that is contaminated by the AR process.

To test the Granger causality, we regress $\{Y_t\}$ on its lagged values alone and on both its lagged values and the lagged values of $\{X_t\}$. If we expect an instantaneous response of $\{Y_t\}$ to $\{X_t\}$, we include the current values of $\{X_t\}$, instead.[13] Finally, we compare the coefficients of determination of the two models with the $F$ test to determine whether any predictive improvement due to the effect of the independent variable is statistically significant.

The concept of the Granger causality was developed in economics. We may apply the concept more flexibly in the social sciences. First, the original Granger causality test is bidirectional. The above procedure is repeated for the effect of $\{Y_t\}$ on $\{X_t\}$. In the social sciences, however, the Granger causality test does not need to be bidirectional. We determine the dependent variable and independent variable(s) according to our theory, not according to the results of statistical analysis.

Second, we often explain trends and seasonality of the dependent variable, rather than eliminating them, in the social sciences. On the other hand, we may have no interest in an AR process of residuals that do not convey important theoretical implications. For example, we may be interested in what causes violent crime rates to increase or decrease, but not in what causes fluctuations in deseasonalized and detrended crime rates. In this case, we test the Granger causal relationship between two nonstationary time series, controlling for the continuity of the dependent variable that is caused by the systematic pattern of residuals. That is, we can try a model with a prewhitened dependent variable and its lagged term along with prewhitened independent variable(s). We will prewhiten variables in our model to prevent the possibility that their coefficients may be over- or underestimated due to the concordance or discordance between their systematic patterns. At the same time, we will include a lagged dependent variable to control the continuity of the dependent variable.

Third, we should consider the characteristics of the dependent variable, rather than statistical tests, to determine whether or not we need to control the continuity of the dependent variable. Many social phenomena tend to be stable, continuing to increase or decrease or remaining high or low for a certain period. However, we may not need to control all these continuities. We will control the continuity of the dependent variable only when continuity or stability is one of the variable's characteristics.

For example, studies generally agree that party identification is stable in many democracies, such as the U.S. (Abramson and Ostrom 1991; Bartels et al. 2010; Campbell et al. 1960; Green et al. 2001), Russia, and the region of Palestine (Abu Sada 1998; Colton 2000; Miller and Klobucar 2000). In this case, we need to include a lagged version of party identification to control the continuity of party identification. In fact, many cross-sectional studies of party identification have included a lagged version of party identification or an instrumental variable (see e.g. Fiorina 1982; Green et al. 2001; Lockerbie 1989, 2002; Markus and Converse 1979; Shin 2012).[14]

On the other hand, in many Latin American and post-communist societies, the political party system is susceptible to various nonparty factors, such as competition among social leaders and religion (Kitschelt 1995; Mainwaring 1999; Mair 1996; Tomsa and Ufen 2013). In this case, party identification itself is not stable. We may observe the same party identification structure for a certain period, but the continuity may be caused by nonparty factors, such as regional sentiment, religion, or social leadership. In this case, we need to deal with the continuity by including these factors in our model. There may be no continuity of party identification itself that we need to control by including its lagged version on the right-hand side of equation.

## B. Cointegration and Error-Correction Model

When the dependent variable and an independent variable contain systematic patterns such as an AR($p$) process, the estimated coefficient of the independent variable may be biased. In addition, the spurious-regression problem may apply (Granger and Newbold 1974). That is, the OLS regression model may produce autocorrelated residuals. The estimated coefficient of the independent variable, its significance test, and the estimated coefficient of determination ($R^2$) may be biased (see chapter 3, section 1A, and chapter 6, section 4).

As explained in section 2, we prewhiten the dependent variable and an independent variable by eliminating their systematic residual patterns for a more accurate estimation of the two variables' relationship. This prewhitening approach assumes that these systematic patterns are simply noise that has nothing to do with the relationship between the two variables. In this approach, we focus on these variables' relationship through their long-term trends. However, systematic residual patterns in the dependent variable and an independent variable may represent their time-dependent behaviors. Therefore, treating such patterns simply as noise may not be appropriate. Time series analysis with prewhitened variables may produce a misleading interpretation of their relationship because prewhitening variables will remove or at least attenuate severely some of the essential information regarding their time-dependent behaviors.

When two variables are *cointegrated*, there exists a valid error-correction mechanism of their relationship and we do not need to prewhiten them to deal with the spurious-regression problem (Engle and Granger 1987). The dependent variable and an independent variable are cointegrated, or in a long-term equilibrium relationship, if they are integrated of the same order—usually of order 1—and share a stochastic process that makes each variable return to its long-term equilibrium status after a sudden change of each variable. The order-$d$ differencing of each of the two variables will leave only a stationary process. For example, two variables are cointegrated if $\{Y_t\}$ and $\{X_t\}$ are *random walks*—$\{Y_{t-1}\}$ or $\{X_{t-1}\}$ plus unpredictable random fluctuations— and therefore if the first differencing will remove unit roots and leave only random fluctuations.

Like a drunken woman and her dog (Murray 1994), cointegrated variables do not drift too far from each other. Even if each of them freely wanders, she can call her dog's name and the dog can bark when they feel that they are too far apart. They will respond to the sound and move closer to each other.[15]

When we plot two cointegrated variables, they will appear to move together in the long run.

Examples of cointegrated variables from Engle and Granger (1987), who originally suggested this concept, are (1) short- and long-term interest rates and (2) per capita consumption and per capita income. We can also assume a cointegrating relationship between capital appropriations and expenditures; household income and expenditures; prices of the same commodity in different markets; and prices of close substitutes in the same market. These are economic indicators. We can easily assume an equilibrium relationship between these indicators. For example, short-term inflation may drift away from long-term inflation during a short time period. However, long-term inflation is the aggregate of short-term inflations. The average of short-term inflation during this period is unlikely to be lower (or higher) than that during another period when long-term inflation is lower (or higher). As a result, both short-term and long-term inflation will appear to trend up and down together.

In the social sciences, we may assume a cointegrating relationship, for example, between presidential approval and support for the incumbent party and between the nuclear capacity of the U.S. and that of the Soviet Union during the Cold War (for more examples, see Clarke and Stewart 1995; Clarke, Stewart, and Whitely 1997; Durr 1993; Rajmaira and Ward 1990; Wlezien 1996).

When two time series, $\{X_t\}$ and $\{Y_t\}$, are cointegrated, $\{X_t\}$ exerts both short-term and long-term influence on $\{Y_t\}$. When a one-unit change in $\{X_t\}$ immediately changes $\{Y_t\}$, the amount of change in $\{Y_t\}$ is the short-term effect of $\{X_t\}$ on $\{Y_t\}$:

$$(9) \qquad (Y_t - Y_{t-1}) = b_0 + b_1(X_{t-1} - X_{t-2}) + e_t$$

In this equation, $b_1$ is the estimated short-term effect of $\{X_t\}$ on $\{Y_t\}$. After the immediate change, $\{Y_t\}$ starts to settle down in the long run to a new equilibrium value according to its own stochastic process. The change in $\{Y_t\}$ due to this settling-down movement is the long-run effect of $\{X_t\}$ on $\{Y_t\}$. This change is not explained by $X_{t-1} - X_{t-2}$. It is reflected in the residuals, $e_t$, instead.

When $\{X_t\}$ and $\{Y_t\}$ are cointegrated, even if $\{X_t\}$ and $\{Y_t\}$ wander from each other, there is a way for them to get back together. The way for them to get back together can be explained by a linear combination of $\{X_t\}$ and $\{Y_t\}$ at lag 1 when $\{Y_t\}$ did not change yet:

(10)
$$Y_{t-1} = a + b_3 X_{t-1} + r_t$$

where $r_t$ is the residuals.

This linear combination characterizes the residuals, $e_t$, in equation (9). With this linear combination, which represents an error-correction mechanism of the equilibrium relationship between $\{X_t\}$ and $\{Y_t\}$, we can rewrite equation (9) as:

(11)     $$(Y_t - Y_{t-1}) = b_0 + b_1(X_{t-1} - X_{t-2}) + b_2(Y_{t-1} + (-b_3)X_{t-1}) + u_t$$

where $u_t$ is the residuals.

In this equation, $b_1$ estimates the immediate effect of the change in $\{X_t\}$ between lag 1 and lag 2 on the change in $\{Y_t\}$ between lag 0 and lag 1. To be accurate, the interpretation of $b_1$ needs to be different from that of the usual OLS regression coefficient, because it represents the relationship between two cointegrating variables. The first differencing will remove unit roots from $\{Y_t\}$ and $\{X_t\}$ and leave only random components. Therefore, $b_1$ explains immediate random disturbances in $\{Y_t\}$ caused by random changes in $\{X_t\}$ at previous lag.

Changes in $\{X_t\}$ disturb the equilibrium between $\{Y_t\}$ and $\{X_t\}$, sending $\{Y_t\}$ on a long-term movement to a value that reproduces the equilibrium state given the new value of $\{X_t\}$. The error-correction portion, $Y_{t-1} + (-b_3)X_{t-1}$, represents this relationship between $\{Y_t\}$ and $\{X_t\}$, which is reflected in the residuals, $e_t$, in equation (9). The coefficient $b_3$ estimates the total long-run effect that a one-unit increase in $\{X_t\}$ has on the change in $\{Y_t\}$ through their long-run equilibrium relationship. This coefficient is also known as the *long-run multiplier*. Increases or decreases in $\{X_t\}$ disturb the long-term equilibrium relationship between these two variables. If $\{X_{t-1}\}$ increases, $\{Y_{t-1}\}$ will become too low ($Y_{t-1} < b_3 X_{t-1}$). Then $\{Y_t\}$ will be adjusted upward by $b_3$ per unit change in $\{X_t\}$ to recover the equilibrium with $\{X_t\}$ in the following period. If $\{X_{t-1}\}$ decreases, $\{Y_{t-1}\}$ will become too high ($Y_{t-1} > b_3 X_{t-1}$). Then $\{Y_t\}$ will be adjusted downward in the following period. Note that $b_3$ is multiplied by $-1$ in the error-correction portion, $Y_{t-1} + (-b_3)X_{t-1}$. We estimate $b_3$ with $Y_t = a + b_3 X_{t-1} + r_t$, but $b_3 X_{t-1}$ is moved to the left-hand side of the equation in the error-correction portion.[16]

The coefficient $b_2$ estimates the error-correction rate, that is, the speed at which $\{Y_t\}$ returns to equilibrium after a deviation caused by a change in $\{X_t\}$ at the previous lag. To put it differently, the total long-term effect of $\{X_t\}$ on

$\{Y_t\}$, $b_3$, will be distributed over future time periods at a rate of $b_2$ per time point until $\{Y_t\}$ and $\{X_t\}$ return to their equilibrium state. If the error-correction approach is appropriate, then $b_2$ will be between $-1$ and $0$. If $b_2$ is positive or if $|b_2| \geq 1$, $\{Y_t\}$ will not return to its equilibrium state with $\{X_t\}$.

For example, if $b_1$, $b_2$, and $b_3$ are 0.5, $-0.2$, and 1.2, respectively, and if $\{Y_t\}$ and $\{X_t\}$ consist of monthly observations, a one-unit increase in $\{X_t\}$ immediately produces a 0.5-unit increase in $\{Y_t\}$. Deviations from equilibrium will be corrected at a rate of 0.2 per month. $\{Y_t\}$ will respond by increasing a total of 1.2 points, spread over future time periods at a rate of 20% per month. $\{Y_t\}$ will increase by 0.24 points at $t$, 0.192 at $t + 1$, 0.154 at $t + 2$, 0.123 at $t + 3$, 0.098 at $t + 4$, 0.08 at $t + 5$, 0.06 at $t + 6$, etc.[17]

For the error-correction model (equation (11)) to work as expected, two conditions should be satisfied. If not, the two variables will not appear to move up and down together in the scatter plots. First, $\{Y_t\}$ and $\{X_t\}$ should be integrated of the same order. Two time series that are integrated of different orders cannot be cointegrated. In the social sciences, the cointegration approach usually assumes that two variables are integrated of order 1. Second, the error-correction portion, $Y_{t-1} - b_3 X_{t-1}$, in equation (11) or the residuals, $r_t$, in equation (10) should be stationary. That is, the coefficient of its AR process should be smaller than 1. In this case, $\{Y_t\}$ will ultimately return to the equilibrium relationship with $\{X_t\}$. If $r_t$ contains a unit root, that is if the coefficient of its AR process is 1, changes in $\{Y_t\}$ caused by changes in $\{X_t\}$ will be permanent, and the variance of $r_t$ will increase with time.

In equation (11), the two coefficients in the error-correction portion, $b_2$ and $b_3$, explain the long-term equilibrium relationship between $\{Y_t\}$ and $\{X_t\}$. The autocorrelation in the residuals, $e_t$, in equation (9) is now explained and thereby controlled out by this error-correction portion. A regression model for the error-correction mechanism, $Y_t = a + b_3 X_{t-1} + r_t$, has autocorrelated residuals—whose AR coefficient in this case is smaller than 1— which will cause a high coefficient of determination and a high level of significance of $b_3$. However, we use $b_3$ only to represent a cointegrating long-run relationship between $\{Y_t\}$ and $\{X_t\}$ that is reflected in the stationary autocorrelated residuals $(r_t)$. In addition, $\{Y_t\}$ and $\{X_t\}$ are differenced in equation (11) so that their unit roots are removed. If our error-correction model is appropriate, the residuals, $u_t$, in equation (11) should be a realization of a white or IID sequence. Therefore, we do not have to worry about the spurious-regression problem, even if $\{Y_t\}$ and $\{X_t\}$ contain systematically patterned residuals.

We can use the Engle-Granger test to test for cointegration between two variables (Engle and Granger 1987). It is a two-step procedure. First, we check whether the dependent variable and an independent variable are integrated of the same order. When the dependent variable and/or an independent variable contain seasonal components, we had better eliminate them before we check whether the dependent variable and an independent variable are integrated of the same order.[18] In most social science studies, the dependent variable and an independent variable are likely to be integrated of order 1. If both variables are integrated of order 1, that is if both variables contain unit roots, then first differencing each variable will successfully remove the unit roots from them. To determine whether the first differencing successfully eliminates a unit root, we need to refer to the sample ACF correlograms, which should decay very fast (see chapter 2, section 5). If the ACF correlograms decay slowly, we may need to try a second differencing. We can also employ the Dickey-Fuller test (see below).

Second, provided that both the dependent variable and an independent variable are integrated of the same order, we regress the dependent variable on the independent variable (see equation (10)). Then we check whether the residuals, $r_t$, contain a unit root. If not, then a linear combination of the two time series, $Y_t - b_3 X_{t-1}$, which is the error-correction portion in equation (11), is stationary. When the residuals, $r_t$, do not contain a unit root, we conclude that the dependent variable and the independent variable are cointegrated. We may try the second procedure first because two variables are not cointegrated, if the residuals we obtain from the regression of $Y$ on $X$ contain a unit root.

We can check whether a variable contains a unit root with various unit-root tests, such as the Dickey-Fuller test (Dickey and Fuller 1979; Fuller 1996; MacKinnon 1996). Briefly, a variable is first-differenced and regressed on its own lagged term:

$$Y_t = \phi Y_{t-1} + e_t$$
$$\nabla_1 Y_t = Y_t - Y_{t-1} = \phi Y_{t-1} - Y_{t-1} + e_t = (\phi - 1)Y_{t-1} + e_t$$

With this model, we test whether the coefficient on the lagged term, $|\phi| - 1$, is significantly different from zero, that is, whether $|\phi|$ is significantly smaller than 1 ($H_0$: $|\phi| - 1 = 0$, that is $|\phi| = 1$; $H_a$: $|\phi| - 1 < 0$, that is $|\phi| < 1$), by referring to the Dicky-Fuller distribution table (see table A2, appendix 4, page 209).[19] The table provides only six sample sizes. As in other statistics, if we do not find a row for our sample size, we go to the next-smallest sample size for a more rigorous test. For a sample size greater than 500, we refer to the last row.

When we fail to reject the null hypothesis, we accept the presence of a unit root in $\{Y_t\}$ and conclude that $\{Y_t\}$ is a random walk. It is more difficult to reject the null hypothesis at $\alpha = .01$ than at $\alpha = .05$. That is, it is more difficult to reject the presence of a unit root at $\alpha = .01$ than at $\alpha = .05$.

For example, the AR coefficient at lag 1 is 0.992 (the Dickey-Fuller $t$ statistic, $-1.962$) for the detrended monthly unemployment rates and 0.987 ($-1.198$) for the detrended monthly inflation. Therefore, we reject the hull hypothesis only for the detrended monthly unemployment rates at $\alpha = .05$. We do not reject the null hypothesis for both variables at $\alpha = .01$ (see table A2). When we regress inflation on unemployment rates, the residuals contain a unit root with 0.977 ($-1.362$) as its AR coefficient at lag 1. Therefore, these two variables are not cointegrated. According to figures 43 and 44, these two variables do not appear to move together. In fact, these two variables are well known for their negative relationship (the Phillips curve). Lower unemployment leads to higher wages, and firms pass the resultant increasing costs on to consumers by increasing prices (Phillips 1958).

In addition, we can visually inspect the ACF correlogram to determine whether a variable contains a unit root. If ACFs decrease slowly even without a deterministic trend, we need to suspect the presence of a unit root. However, we need to be careful in concluding, from the visual inspection, that a unit root is present. If $|\phi|$ is quite strong, ACFs are likely to show a slow decrease even when $|\phi|$ is significantly smaller than 1. Thus, when the presence of a unit root is suspected, we need to conduct both visual inspection and statistical tests.

The Dickey-Fuller test does not distinguish between true unit-root processes ($\phi = 1$) and near or fractional unit-root processes. However, many social science time series are not consistent with unit-root processes (Beck 1992; Williams 1992). Many variables are near or fractionally integrated.[20] It may be inappropriate to consider only the unit root to identify cointegration in most social science studies because integration and cointegration may be a matter of degree. The effect of a shock on a time series with a unit-root process never dies out. The effect of a shock on a stationary series dies out quickly. The effect of a shock on a near or fractionally integrated time series persists for long time but will die out eventually.[21] When we deal with a near or fractional unit-root process, we discuss an integration order between 0 and 1, not 0 or 1.

Cointegration may exist between the dependent variable and two or more independent variables. Equation (11) can be extended to include more cointegrating independent variables:

$$(12) \quad (Y_t - Y_{t-1}) = b_0 + b_{1a}(X_{a,t-1} - X_{a,t-2}) + b_{1b}(X_{b,t-1} - X_{b,t-2}) + b_{1c}(X_{c,t-1} - X_{c,t-2}) + \ldots + b_2(Y_{t-1} - b_{3a}X_{a,t-1} - b_{3b}X_{b,t-1} - b_{3c}X_{c,t-1} - \ldots) + e_t$$

In this extended model, each independent variable is forced to make an independent contribution to the long-term relationship. The Engle-Granger test for equation (11) will not tell us which independent variables have cointegrating relationships with the dependent variable. In this case, we can employ the Johansen (1988) test. Alternatively, we first employ the Engle-Granger test for equation (11). This test will tell us whether cointegration is present. Then we employ the Engle-Granger test for each independent variable, which will tell us whether the dependent variable has a cointegrating relationship with each independent variable.

In sum, to model a cointegrating relationship between $\{X_t\}$ and $\{Y_t\}$, we first check whether two variables are integrated of the same order and whether the residuals from the regression of $Y$ on $X$ are stationary. Then we regress changes of $Y$ at the current lag, $Y_t - Y_{t-1}$, on changes of $X$ at lag 1, $X_{t-1} - X_{t-2}$, and the lagged residuals, $Y_{t-1} - b_3X_{t-1}$.[22] With this model, we can estimate the relationship between $\{X_t\}$ and $\{Y_t\}$ that is not influenced by the spurious-regression problem.

Several points deserve attention when we apply the idea of cointegration to social science studies. First, cointegration is a statistical concept, rather than a theoretical concept. The long-run equilibrium relationship between two variables does not need to come from their causal relationship. It can simply be a relationship between similarly behaving variables. In the social sciences, however, we need to discuss a cointegrating relationship between two variables with caution. We are usually required to theoretically explain why we expect a particular relationship between two variables, rather than simply to report the relationship we observe.

Even when we observe that two variables tend to move together in the social sciences, we may not theoretically assume cointegration, an equilibrium relationship between the two variables. For example, even if presidential popularity (Ostrom and Smith 1992; Clarke and Stewart 1996) or the violent crime rates that are analyzed in this book tend to move together with economic conditions, we may not be sure about whether they are bound to move together. The tendency for two variables to move together may be limited to the period of observation. Various factors may influence presidential popularity and violent crime rates. Presidential popularity may remain high even when economic conditions get worse. Violent crime rates may remain high even when economic conditions get better. In addition, the tendency for two variables to move together may be limited regionally. For example, economic adversity is expected to lead to increasing violent crime rates in advanced democracies. However, as developing countries prosper more and more, they also become increasingly modernized, and violent crime rates are likely to increase, even with economic prosperity. Therefore, in the social sciences, we may not theoretically generalize a cointegrating relationship between two variables even when visual inspection and statistical tests reveal such a relationship.

Second, cointegration assumes that variables with a cointegrating relationship are jointly endogenous. If we assume a cointegrating relationship between $\{Y_t\}$ and $\{X_t\}$, $\{Y_t\}$ can cause changes in $\{X_t\}$ as well as $\{X_t\}$ can cause changes in $\{Y_t\}$. However, we clearly set the dependent variable and independent variable(s) according to our theory in most social science studies. In fact, the relationship between two variables may not be recursive in the social sciences. We may identify causality from one of them to the other, but not vice versa. For example, economic conditions influence macropartisanship (Mackuen et al. 1989; Weisberg and Smith 1990) and individual voters' party identification (Lockerbie 1989, 2002), but partisanship does not influence economic conditions or shape economic perceptions. In this case, even if we observe that two variables tend to move together in the long run, we cannot assume a cointegrating relationship between them.

Third, observations at different time points are often discrete and independent of each other in the social sciences. Most economic indicators and natural phenomena are continuous, even when they are measured per a fixed time interval. We know that they continue to move between every two time points. On the other hand, some social science time series data consists of actual discrete temporal observations that are independent of each other.[23] They are time series data in that they are collected across time through repeated regular temporal observations on a single unit of analysis and entered in chronological order. However, there is no such movement to return to the equilibrium relationship with another time series. In this case, we should not assume a cointegrating relationship between two time series variables, even when they satisfy the statistical conditions for cointegration.

For example, the reported vote percentage of the winning candidates in congressional elections may appear to move together with the actual vote percentage for the candidates. In the U.S. congressional elections, where more than 90% of representatives and more than 80% of senators tend to be reelected with safe margins, the reported and actual vote percentages for the winning candidates are likely to remain stable. That is, they are likely to contain something akin to a unit-root process. However, each election is a separate independent event. Election outcomes can be measured only every two years in the U.S., and they do not exist or continue to change during the two-year period. Votes for the winning candidates tend to be over-reported in proportion to their actual vote percentages (see e.g. Eubank and Gow 1983; Gronke 1992; Wright 1990, 1992). However, we cannot assume a mechanism through which actual vote percentages and over-reported vote percentages recover their equilibrium relationship.

## C. Long-Memory Effect and Autoregressive Distributed Lags Model

Variables are more likely to contain a long-memory behavior than a permanent-memory behavior (a unit-root process) in the social sciences. They may respond to shocks far in the past, but such a response will become weaker over time. For example, $AR(p)$ processes usually die off exponentially and therefore quickly (a short-memory process), but $AR(p)$ processes with high coefficient, $\phi$, do not die off quickly and may persist quite long (a long-memory process). In the latter case, the effect of an independent variable on the dependent variable will not be maintained permanently, but it will take time for the effect to virtually disappear.

As suggested in chapter 2, we usually do not difference our time series to make it stationary in the social sciences. We estimate and then eliminate systematic patterns from our time series. Therefore, by the time we are ready to run an analysis to test our hypothesis, we will know whether each variable contains a stochastic process that persists quite long or dies off quickly. We will also know its integration order. Then we need to determine whether each variable's stochastic process that persists quite long produces a long-memory relationship between $\{X_t\}$ and $\{Y_t\}$. We may determine that each variable's systematically patterned residuals are simply noise that has to be eliminated for a more accurate estimate of the effect of $\{X_t\}$ on $\{Y_t\}$ (see section 2 above). If we theoretically assume a long-memory effect of $\{X_t\}$ on $\{Y_t\}$, we need to incorporate a component that captures this long-memory effect in our model. When $\{Y_t\}$ and $\{X_t\}$ are not cointegrated but $\{X_t\}$ has a long-memory effect on $\{Y_t\}$, we can utilize the following *autoregressive distributed lags* model, ADL $(p, n)$, in which $\{Y_t\}$ is a function of past lags of $\{Y_t\}$ and current and past lags of $\{X_t\}$:

$$Y_t = b_0 + b_{y1}Y_{t-1} + b_{y2}Y_{t-2} + \ldots + b_{yp}Y_{t-p}$$
$$+ b_{x1}X_t + b_{x2}X_{t-1} + b_{x3}X_{t-2} + \ldots + b_{xq}X_{t-(q-1)} + e_t$$

This model consists of two components. One is the AR process of $\{Y_t\}$ and is represented by the lagged dependent variable(s) on the right-hand side. Current values of the dependent variable may be a function of its prior values at multiple lags. Therefore, we may need to include more than one lagged dependent variable, representing lag 1 to lag $p$ previous values of the dependent variable, respectively. The other is the distributed lags of $\{X_t\}$, which deal with the current and lagged effects of $\{X_t\}$ on $\{Y_t\}$. In this model, $b_{x1}$ estimates the immediate effect of $\{X_t\}$ on $\{Y_t\}$, and $q - 1$ is the lag length during which the effect of $\{X_t\}$ on $\{Y_t\}$ persists. The sum of the coefficients of all the current and lagged terms of $\{X_t\}$ estimates the long-memory effect of $\{X_t\}$ on $\{Y_t\}$ distributed over time. If any of $b_x$ coefficients and any $b_y$ are significant, an ADL model is a Granger cause model (see equation (8), page 143).

The ADL (1, 1) model that includes the first lagged dependent variable, $\{Y_{t-1}\}$, and an independent variable whose lag length is 1—$\{X_t\}$ and $\{X_{t-1}\}$—is isomorphic to the error-correction model, and it can be rewritten in the error-correction form (see equation 11):

(13)    $Y_t = b_0 + b_2Y_{t-1} + b_1X_t + b_3X_{t-1} + e_t$

$(Y_t - Y_{t-1}) = b_0 + b_1X_t - b_1X_{t-1} + b_1X_{t-1} + b_2Y_{t-1} - Y_{t-1} + b_3X_{t-1} + e_t$

(14)    $= b_0 + b_1(X_t - X_{t-1}) + (b_1 + b_3)X_{t-1} + (b_2 - 1)Y_{t-1} + e_t$

(15)    $= b_0 + b_1(X_t - X_{t-1}) + (b_2 - 1)\{Y_{t-1} + \dfrac{(b_1 + b_3)}{(b_2 - 1)}X_{t-1}\} + e_t$

Transforming an ADL model into an error-correction model allows us the error-correction type of interpretation of the relationship between $\{Y_t\}$ and $\{X_t\}$ (see section 3B above), although the two variables are not cointegrated. In equations (14) and (15), $b_1$ is an immediate (short-term) effect and $\dfrac{(b_1 + b_3)}{(b_2 - 1)}$

is a total long-run effect (long-run multiplier) that changes in $\{X_t\}$ have on changes in $\{Y_t\}$. The coefficient $b_2$ is the proportion of the deviation of $\{Y_t\}$ at the previous lag that is maintained at the current lag, and $b_2 - 1$ is the speed at which $\{Y_t\}$ corrects its deviation in each subsequent time period. The amount of change in $\{Y_t\}$ at the current lag caused by the change in $\{X_t\}$ at the previous lag will be $b_1 + b_3$ in equation (14) and $\dfrac{(b_1 + b_3)}{(b_2 - 1)} \times (b_2 - 1)$ in equation (15). As in the error-correction model, $b_2 - 1$ will be between $-1$ and $0$, and $|b_2| < 1$, if the error-correction model approach is appropriate.

The long-run multiplier, $\dfrac{(b_1 + b_3)}{(b_2 - 1)}$, is not estimated in equation (15). Thus, its standard error needs to be calculated. The standard error of $\dfrac{(b_1 + b_3)}{(b_2 - 1)}$ can be determined through the Bewley transformation regression (Bewley 1979; De Boef and Keele 2008):

(16) $$Y_t = b_0 + b_1(Y_t - Y_{t-1}) + b_2 X_t + b_3(X_t - X_{t-1}) + e_t$$

In this equation, $b_2$ is identical to $\dfrac{(b_1 + b_3)}{(b_2 - 1)}$, and the standard error of $b_2$ is the standard error of $\dfrac{(b_1 + b_3)}{(b_2 - 1)}$. Since we have the dependent variable, $Y_t$, on the right-hand side, we first obtain the predicted values of $Y_t - Y_{t-1}$ with equation (14). Then, using the predicted values of $Y_t - Y_{t-1}$ for $Y_t - Y_{t-1}$, we estimate equation (16).

Two points draw our attention with regard to the ADL model. First, equations (14) and (15) are not error-correction models between two cointegrated variables. Therefore, our model may have a spurious-regression problem, unless $\{e_t\}$ in equation (13) is a realization of white noise. In equations (14) and (15), we add $Y_{t-1}$ to both sides of the equation, and we add and then subtract $X_{t-1}$ on the right-hand side. Therefore, the transformation from equation (13) to equations (14) and (15) does not change the characteristics of $\{e_t\}$. We should not prewhiten $\{Y_t\}$ and $\{X_t\}$ to make $\{e_t\}$ a realization of white noise. A main purpose of utilizing the error-correction mechanism is to avoid prewhitening our variables and to keep the information of their systematic residual patterns, while avoiding a spurious-regression problem. It does not make sense to prewhiten $\{Y_t\}$ and $\{X_t\}$ and thereby to eliminate the information of their long-run relationship and then to apply the error-correction mechanism to estimate their relationship.

Second, we can assume an ADL model only if $\{Y_t\}$ contains an AR process but not a deterministic trend. Otherwise, $\{Y_{t-1}\}$ is not an AR term, and therefore it is not appropriate to call equation (13) an ADL model. It may be a Granger cause model, if the coefficients of $\{X_t\}$ and/or its lagged terms are significant. In addition, ADL models are OLS regression models that employ the lagged dependent and independent variables on the right-hand side of the equation. Even when $\{Y_t\}$ contains only an AR process, the coefficient of $\{Y_{t-1}\}$ may be different from that of a pure AR process, if $\{X_t\}$ and its lagged terms are correlated with $\{Y_{t-1}\}$ (see Section 3-A).

By transforming the ADL model into the error-correction model (equations (14) and (15)), we are likely to lessen high multicollinearity problems in equation (13). In equation (13), an independent variable and its lagged term, that is the same independent variable shifted backwards, are included. These two variables are likely to be highly correlated with each other, especially when the independent variable contains systematic patterns. On the other hand, the independent variable is replaced with its first-differenced form in equations (14) and (15). If the independent variable contains a time-dependent behavior, its first-differenced values will be significantly different from its original values and from values of its lagged term. Therefore, the correlation between the lagged independent variable and the first-differenced independent variable will be lower than that between the independent variable and its lagged term.

## D. The Case of Monthly Violent Crime Rates, 1983–1992

*Granger Causality.* Crime rates are different from affective attitudes, such as party identification and presidential popularity. There is no reason for which we expect crime rates to be stable, independent of various factors. We do not expect a kind of built-in inertia of crime rates, either. Violent crime rates may be high or low for a certain period, but such high or low crime rates will be produced by various factors. Violent crime rates are likely to decrease with improved economic conditions. Potential criminals may not commit crimes, and habitual criminals are less likely to commit crimes, when their economic adversity is lessened. If recidivists significantly increase crime rates, the implementation of a getting-tougher-on-recidivists policy will significantly reduce crime rates.

Since we do not assume continuity as a characteristic of the monthly violent crime rates, we may not need to include a lagged version of crime rates

TABLE 11. Monthly Economic Adversity and Monthly Violent Crime Rates, 1983–1992

| | Lagged D.V. | | Lagged I.V. | Lagged D.V. & I.V. |
|---|---|---|---|---|
| | Original variables | Prewhitened variables | | |
| Violent crime rates at lag 1 | .103* (.045) | .013 (.086) | | .022 (.089) |
| Unemployment rate | .077 (.157) | −.092 (.128) | .464 (.542) | .456 (.546) |
| Unemployment rate at lag 1 | | | −.55 (.538) | −.552 (.54) |
| Consumer price index | .44*** (.028) | .486*** (.044) | .646* (.388) | .66* (.393) |
| Consumer price index at lag 1 | | | −.157 (.388) | −.182 (.402) |
| Constant | −4.79 (2.45) | −4.26* (2.01) | −3.97* (2.02) | −3.87* (2.07) |
| $N$ | 119 | 119 | 119 | 119 |
| Adj. $R^2$ | .92 | .959 | .959 | .959 |

$*p < .05, ***p < .001$ (one-tailed test).
NOTE. Numbers in parentheses are standard errors.

on the right-hand side. However, we may consider the fact that violent crime rates continued to increase between 1983 and 1992. Such a continuous increase may be caused not only by economic adversity but also by other factors, such as low costs for committing crimes. Therefore, we may include a lagged dependent variable whose coefficient will reflect the effect of excluded important independent variables on the dependent variable.

Table 11 reports a model with a lagged dependent variable on the right-hand side. The first and second columns include a lagged dependent variable.[24] Column 3 includes lagged independent variables but not a lagged dependent variable. Column 4 includes a lagged dependent variable along with lagged independent variables. Variables are not whitened for the first column but are whitened for the other columns. The dependent variable and its lagged term were deseasonalized.[25]

As expected, in the model in which the dependent variable and its lagged term were not whitened (column 1), the lagged term's coefficient is significant and positive. Inflation is statistically significant, but its coefficient is reduced, compared with its coefficient in table 9.

Since we did not prewhiten the crime rates, the dependent variable and its lagged term are significantly correlated with each other through their identical AR process. In addition, the lagged term may be correlated with residuals, which are the differences between the observed and predicted values of the dependent variable. The residuals contain an AR(2) process.

In column 2, we prewhitened the dependent variable and thereby eliminated the possibility that the dependent variable and its lagged term may be autocorrelated. We see that the coefficient of the lagged dependent variable is no longer significant, supporting our argument that we do not expect continuity in monthly violent crime rates. The coefficients of prewhitened unemployment rate and prewhitened inflation are similar to those in the first column of table 10, not requiring significantly different interpretation.

According to the second, third, and fourth columns, the lagged dependent variable and the lagged independent variables do not significantly contribute to the explanation of the dependent variable. According to the adjusted $R^2$, the amount of the explained variation of the dependent variable remains identical to that of the model that does not include these lagged variables (see column 1 in table 10). In addition, the level of significance for inflation reduces in the third and fourth columns, compared with table 10. This reduced level of statistical significance is likely to be caused by extremely high collinearities, especially between the original and lagged independent variables. Bivariate correlations between prewhitened unemployment and its lagged term and between prewhitened inflation and its lagged term approach 1. As mentioned above, it may not be appropriate to include the lagged terms of unemployment rates and inflation in the model (see section 1D above). In addition, the lagged variables do not statistically contribute to the explanation of the dependent variable, while causing very serious multicollinearity problems. The lagged dependent variable does not significantly explain the current dependent variable, once the dependent variable is prewhitened. Thus, in the case of the monthly violent crime rates, it seems that we had better focus on the crime rate's instantaneous response to economic adversity (column 1 in table 10).

In the social sciences, however, there are cases in which we need to additionally include a lagged dependent variable and a lagged independent variable to accurately explain the independent variable's relationship with the dependent variable. For example, we can expect built-in inertia in affective attitudes, such as party identification and presidential popularity. Many voters may remain loyal to a political party whether they evaluate the party positively or negatively. Many voters may continue to approve the president's handling of his job regardless of intervening events. In these cases, we had better include a lagged dependent variable on the right-hand side of our equation to control the continuity of the dependent variable (see e.g. Fiorina 1982; Kernell 1978; Lockerbie 1989, 2002; Markus and Converse 1979; Shin 2012). In addition, the dependent variable may respond not only to the current values of an independent variable but also to its past values. For example, voters may remember past performances of the president or past economic conditions and base their evaluation of the president on such memories (Beck 1991).

*Error-Correction Model and ADL Model.* We prewhitened monthly violent crime rates and inflation in tables 10 and 11 because we treated each variable's systematic residual pattern as noise that has to be eliminated to more accurately estimate the two variables' relationship. We may treat the two variables' systematic residual patterns differently as those which convey important information of a long-term relationship between the two variables.

We cannot confidently assume a cointegrating relationship between crime rates and inflation. Their relationship is clearly different from that between short-term inflation and long-term inflation. Factors other than inflation may influence crime rates, and as a result, crime rates may drift far away from inflation. In fact, crime rates and inflation appeared to move together in the U.S. until October 1994, but they have drifted away from each other since then. Monthly violent crime rates began decreasing since October 1994 with the implementation of a tougher crime-control policy (see chapter 7). On the other hand, consumer price index, adjusted with 1982–1984 as the base period, continued to increase almost linearly. Here, we check their cointegrating relationship just for discussion.

To check their cointegrating relationship, we need to eliminate their trend and/or seasonality, which will mask their long-term relationship through their stationary stochastic processes. According to figure 52, these two variables may cointegrate during the period from 1983 to 1992. During this period, they tend to move in the same direction, although crime rates frequently drift away from inflation.

The two variables do not pass the Engle-Granger test for cointegration. The deseasonalized and detrended monthly crime rates and the detrended monthly consumer price index from 1983 to 1992 satisfy only the second condition for cointegration. The residuals from the regression of the deseasonalized and detrended crime rates on the detrended consumer price index contain a stationary process that is characterized by an AR(2) process

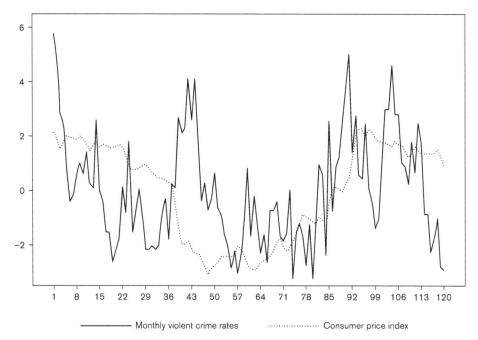

——————— Monthly violent crime rates    ················· Consumer price index

FIGURE 52. Deseasonalized and detrended monthly violent crime rates and detrended consumer price index, 1983–1992.

(figure 53). The AR coefficient is 0.542 at lag 1 and 0.255 at lag 2.[26] In the Dickey-Fuller test, the coefficient of the lagged residuals is significantly different from zero (−0.33). Its $t$ statistic (−4.979) is greater than the critical value in table A2 (−1.95, $\alpha = .05$, $n = 100$). Therefore, we reject the null hypothesis and conclude that a linear combination of crime rates and inflation that produces stationary residuals exists.

However, the two variables do not satisfy the first condition for cointegration: they are not integrated of the same order. The deseasonalized and detrended monthly crime rates contain an AR(2) process (see chapter 2). The detrended monthly consumer price index contains a unit root at $\alpha = .01$, but not at $\alpha = .05$. Therefore, the two variables do not statistically satisfy the first condition, and we do not need to try an error-correction model for these two variables.

We may assume a long-memory relationship between these two variables as their systematically patterned residuals tend to move in the same direction. Inflation immediately increased violent crime rates (see column 1 in table 10). Then, its effect may persist for a long time. Therefore, instead of simply

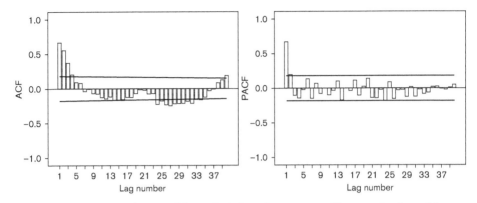

FIGURE 53. ACF and PACF of the residuals from the regression of deseasonalized monthly violent crime rates on consumer price index, lags 1–40.

treating the systematically patterned residuals of each variable as noise and estimating only the immediate effect of inflation on crime rates with pre-whitened crime rates and inflation, we may try an ADL model.

Table 12 reports the ADL models expressed in the form of equation (14) and the error-correction model (equation (11)). The latter is identical to the ADL model expressed in the form of (15). The monthly violent crime rates were deseasonalized. The seasonal components mask the relationship between crime rates and inflation. We can simply add or multiply them back later to the predicted crime rates (see chapter 2). We did not detrend the two variables because their trends are the most important parts in our explanation of their relationship. Deterministic trends are often of major concern in the social sciences, and removing them just to build an ADL model may not be appropriate. Therefore, what we try in table 12 is not an ADL model. Instead, we utilize the idea of the ADL model for the error-correction type of interpretation of the long-memory relationship between the two variables.

According to the model expressed in the form of equation (14), inflation's immediate effect on crime rates is 1.128. The coefficient of the lagged dependent variable indicates that the crime rate corrects its deviation in each subsequent time period by a rate of −0.462. In other words, the deviation of $\{Y_t\}$ at the previous lag is maintained at the current lag by a rate of 0.538 ($b_2 - 1 = -0.462$). The amount of change in crime rates at current lag caused by change in inflation at the previous lag is 0.219. The total long-run effect (long-run multiplier) that changes in inflation have on changes in crime rates is 0.474

TABLE 12. Autoregressive Distributive Lags Model and Error Correction Model for Monthly Economic Adversity and Monthly Violent Crime Rates, 1983–1992

|  | Coefficient | Standard error |
|---|---|---|
| Autoregressive distributive lags model |  |  |
| Consumer price index – Consumer price index at lag 1 | 1.218* | .546 |
| Consumer price index at lag 1 | .219*** | .038 |
| Violent crime rates at lag 1 | –.462*** | .073 |
| Constant | –1.497 | 1.407 |
| $N$ | 119 |  |
| Adj. $R^2$ | .276 |  |
|  |  |  |
| Error correction model |  |  |
| 1. Violent crime rates at lag 1 = $b_3$(consumer price index at lag 1) | .502*** | .014 |
| 2. Consumer price index – consumer price index at lag 1 | 1.157* | .544 |
| Violent crime rates at lag 1 – $b_3$(consumer price index at lag 1) | –.463*** | .073 |
| Constant | –.214 | .255 |
| $N$ | 119 |  |
| Adj. $R^2$ | .274 |  |

*$p < .05$, ***$p < .001$ (one-tailed test).

(0.219/0.462). The standard error for this total long-run effect, estimated with the Bewley transformation regression, is 0.012.[27]

According to the error-correction form (equation (15)), the immediate effect of changes in inflation on changes in crime rates is 1.157 ($\approx$1.128). The deviation of $\{Y_t\}$ at the previous lag is maintained at the current lag by a rate of 0.537 ($\approx$ 0.538). The crime rate corrects its deviation in each subsequent time period by a rate of –0.463 ($\approx$ –0.462). The amount of change in crime rate at the current lag caused by change in inflation at the previous lag is 0.232 (–0.502 × –0.463), which is very close to 0.219. The total long-run effect that changes in inflation have on changes in crime rate is 0.502 ($\approx$ 0.474).

We did not prewhiten the two variables. However, both models do not produce autocorrelated residuals. Only two ACFs and PACFs cross the confidence bounds (figure 54). By incorporating the error-correction mechanism on the right-hand side, we controlled out the AR(2) process from the residuals produced by regressing the deseasonalized monthly violent crime rates on inflation (figure 53).

When we detrended both variables, only the lagged dependent variable in equation (14) and only the error-correction portion in equation (15) are significant. Both terms have identical coefficients (–0.329, standard error 0.067).

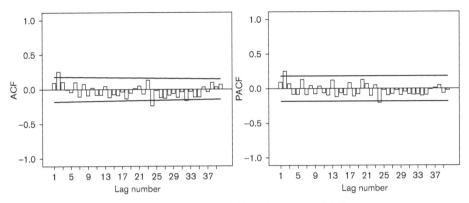

FIGURE 54. ACF and PACF of the residuals from the ADL models, lags 1–40.

This is the expected result, in that we eliminated the portion of the crime rates that we expect inflation to explain and the portion of the inflation that we expect to explain the crime rates. The systematic residual processes of inflation and its lagged term do not significantly influence the systematic residual process of monthly crime rates.

## (4) DIAGNOSTICS

In bivariate or multiple time series analysis, residuals represent the dependent variable's variation that is not explained by independent variable(s). As in OLS regression analysis of cross-sectional data and in univariate time series analysis, several properties of residuals should be satisfied for estimated coefficients to be robust:

(a) a normal distribution
(b) a constant variance, $\sigma^2$, for all the values of the independent variable
(c) a zero mean
(d) no autocorrelation.

In univariate time series analysis, we do not need to check most of these properties, except the normality of residual distribution. When we estimate and eliminate trend, seasonality, and systematic pattern of residuals, we take care of violations of a zero mean, a constant variance, and no autocorrelation (see chapters 2 and 3). In bivariate or multiple time series analysis, we do not decompose the dependent variable and independent variable(s). Instead, we explain the dependent variable's time-dependent behavior in terms of its relationship with one or more independent variables, as we do in OLS regression analysis of cross-sectional data. Therefore, these properties are not taken care of during the model estimation process.

Among these properties of residuals, we usually need to worry most about no autocorrelation in residuals, which is an important property for the robustness of our coefficient estimates. Multiple or bivariate time series analysis is more likely to produce autocorrelated residuals than OLS regression analysis of cross-sectional data.

Autocorrelation in residuals of a multiple time series model can be caused by various factors. It may be caused by systematic patterns of the dependent variable and independent variables. As in OLS regression analysis of cross-sectional data, autocorrelation may be caused by an incorrect functional form of an independent variable. Autocorrelation may reflect the influence of excluded independent variables on the dependent variable.

Therefore, whether or not we prewhitened our variables, we need to conduct visual inspection and randomness tests to check that the residuals of our model do not show any temporal regularity. As explained in chapter 2, visual inspection of the scatter plots and sample ACF and PACF correlograms of residuals is generally more appropriate than statistical tests. We can refer to statistical tests to confirm what we find from our visual inspection of residuals.

We can employ all the randomness tests that were explained in chapter 2, except the Durbin-Watson $d$ test statistic. When we include a lagged dependent variable as an independent variable, the Durbin-Watson $d$ test statistic is biased toward 2. Consequently, even if serial correlation is present, we may fail to reject the null hypothesis that residuals are a realization of IID (or at least white) noise. We need to replace the Durbin-Watson $d$ test with alternative tests, such as Durbin's $h$ test, for the lagged dependent variable (LDV) model. The formula of Durbin's $h$ test is

$$h = \left(1 - \frac{d}{2}\right)\sqrt{\frac{n}{1 - n\text{Var}(b_{LDV})}}$$

where $d$ is the Durbin-Watson $d$ statistic, $n$ is the number of observations, and $\text{Var}(b_{LDV})$ is the variance of the coefficient of the lagged dependent variable.

If programs do not report Durbin's $h$ test statistic, we can calculate the Durbin's $h$ test statistic because we will know $d$ and $\text{Var}(b_{LDV})$ with most statistical programs. In large samples, the Durbin's $h$ test statistic has a standardized normal distribution. We reject the null hypothesis that the coefficient of an AR(1) process is not significantly different from 0, when $|h|$ is greater than the critical value of the $z$ score. Like the Durbin-Watson $d$ test, Durbin's $h$ test can be used only for an AR(1) process. In addition, if $1 - n\text{Var}(b_{LDV})$ is negative, the Durbin's $h$ test statistic cannot be calculated.

We can check whether the order of autocorrelation in the residuals is higher than 1 through our visual inspection of the residuals. Then, we need to replace the Durbin-Watson $d$ test or Durbin's $h$ test with the Breusch-Godfrey test (Breusch 1978; Godfrey 1978). The Breusch-Godfrey test can be employed more generally than the Durbin-Watson $d$ test and Durbin's $h$ test for any AR($p$) processes ($p \geq 1$) and whether or not our model includes lagged dependent variables. We regress the error term, $\{E_t\}$, on its $p$ lagged values as well as the independent variables in the model:

$$E_t = b_0 + b_1 X_{1t} + b_2 X_{2t} + \ldots + \phi E_{t-1} + \phi^2 E_{t-2} + \ldots + \phi^p E_{t-p} + Z_t$$

With the $R^2$ for this model, we calculate the Breusch-Godfrey statistic:

$$(n - p)R^2$$

where $n$ is the number of observations and $p$ is the number of lagged error terms.

The distribution of this statistic is approximated by $\chi^2_{1-\alpha}$ with $p$ degrees of freedom. If the level of significance of the Breusch-Godfrey statistic is less than .05, we reject the null hypothesis that all the coefficients of the AR($p$) process are not significantly different from zero. Namely, we conclude that there is serial correlation of any order up to $p$. Alternatively, we can use the $F$ test for the joint significance of all the coefficients of an AR($p$) process. If $F$ is greater than the critical $F$ value, we reject the null hypothesis.

Autocorrelated residuals violate an important assumption of independent errors in OLS regression analysis. With autocorrelated residuals, the estimated coefficients may not be BLUE (best linear unbiased estimate), as in OLS regression analysis of cross-sectional data. For example, we may overestimate coefficients with the presence of a positive autocorrelation, although they are unbiased in the long run (Berry 1993). In addition, estimated standard errors of coefficients are likely to be biased. For example, the standard errors of coefficients may be underestimated in the presence of a positive autocorrelation.

In bivariate or multiple time series analysis, we can deal with autocorrelated residuals in several ways. First, if autocorrelation in residuals is caused by systematic patterns of the dependent variable and/or independent variables that are not cointegrated, prewhitening these variables will make residuals no longer autocorrelated. For example, autocorrelation in residuals (table 9, figure 41) disappeared after we prewhitened the variables (column 1 in table 10, figure 51).

Second, residuals may be autocorrelated even when we prewhitened variables or when our variables do not contain systematic patterns. In this case, we need to consider the possibility that autocorrelated residuals may reflect the influence of excluded independent variables on the dependent variable. If we can identify excluded variables, we will revise our model to reflect their relationship with the dependent variable adequately. However, it is usually difficult to pinpoint the excluded independent variable(s) in the social sciences. In this case, autocorrelation in residuals may disappear after we include a lagged dependent variable because the effect of any excluded independent variables will be reflected in the lagged dependent variable. For example, the AR(2) process in the residuals produced by regressing the deseasonalized monthly violent crime rates on inflation (figure 53) disappeared when we included the lagged dependent variable and the lagged independent variable on the right-hand side (figure 54). If we observe autocorrelated residuals with a lagged dependent variable on the right-hand side, we need to suspect a misspecification of our lagged model. That is, we may need to include more lagged dependent variables, as long as we can theoretically justify their inclusion in our model. We need to be careful in including lagged dependent variable(s) on the right-hand side to solve the autocorrelated-residuals problem. We transform our model into the Granger causality model (see section 3A) by including a lagged dependent variable. If the AR process in the residuals is simply noise and if we do not assume the continuity of the dependent variable, including a lagged dependent variable just to control out the AR process in residuals may cause a more serious problem, namely a specification error.

Third, systematic patterns in residuals may be caused by incorrect functional forms of the relationship between the dependent variable and an independent variable. For example, we may specify a linear form of the relationship between the dependent variable and an independent variable, when the true relationship is nonlinear. In this case, transforming an independent variable to consider a different functional form of the relationship may solve the autocorrelated-residuals problem. However, when we transform the independent variable, its estimated coefficient will have least square properties with respect to the transformed values and not to the original values. This may be okay in the natural sciences and the statistics. In the social sciences, however, our interpretation of the coefficient will not contain an intuitive meaning of the relationship between the independent variable and the dependent variable.

Finally, we may decide to estimate the AR process of residuals and include it in our model, rather than eliminate it for more accurate model estimation. When we include the AR process of residuals in our model, however, estimated regression coefficients may not be BLUE. Thus, we need to adjust estimated coefficients when we include the AR process of residuals in our model. Some programs, such as SAS, SPSS, and R, allow us to specify a model with residuals that contain an AR process. Even when programs do not allow us to specify such a model, we can estimate adjusted regression coefficients and their standard errors, following the procedure proposed by Cochrane and Orcutt (1949). In this procedure, the model is assumed to contain an AR($p$) process, which represents a constant rate at which the effect of each independent variable dies off per every time interval.[28] The basic steps are:

1. Run a bivariate or multiple time series analysis to estimate the model:

$$Y_t = b_0 + b_1 X_{1t} + b_2 X_{2t} + \ldots + E_t,$$

where $X_{1t}, X_{2t}$ are independent variables and $E_t$ is an error term.

2. Check whether the residuals, $\{E_t\}$, contain an AR($p$) process, and if so, estimate the coefficient(s) of the process (see chapter 2):

$$E_t = \phi E_{t-1} + \phi_2 E_{t-2} + \ldots + \phi_p E_{t-p} + Z_t,$$

where $Z_t$ is white noise.

3. Transform the dependent variable and independent variable(s) with the estimated AR coefficient(s):

$$\hat{Y}_t = Y_t - \phi_1 Y_{t-1} - \ldots - \phi_p Y_{t-p}$$

$$\hat{X}_{1,t} = X_{1,t} - \phi_1 X_{1,t-1} - \ldots - \phi_p X_{1,t-p}$$

$$\vdots$$

4. Rerun a bivariate or multiple time series analysis with the transformed variables to estimate the model:

$$\hat{Y}_t = \beta_0 + \beta_1 \hat{X}_{1,t} + \beta_2 \hat{X}_{2,t} + E_t$$

where $E_t = \phi E_{t-1} + \phi_2 E_{t-2} + \ldots + \phi_p E_{t-p} + Z_t$.

The estimated coefficients $\beta_1, \beta_2, \ldots$ are adjusted estimates of the corresponding coefficients of the original model. Their standard errors are likely to increase if the estimated AR coefficient is positive because the standard errors tend to be underestimated with the positively autocorrelated residuals (see chapter 3). The adjusted intercept will be $\beta_0/(1 - \phi_1 - \ldots - \phi_p)$, with the standard error equal to (standard error of $\beta_0$)/$(1 - \phi_1 - \ldots - \phi_p)$. The error term, $\{Z_t\}$, will be a realization of white noise.

A weakness of the Cochrane-Orcutt procedure is that the first $p$ observation(s) become unavailable. The Prais-Winsten (1954) procedure makes a reasonable transformation for the first observation in the following form:

$$\hat{Y}_1 = \sqrt{1-\phi^2} \times Y_1$$

$$\hat{X}_{1,1} = \sqrt{1-\phi^2} \times X_{1,1}$$

$$\hat{X}_{2,1} = \sqrt{1-\phi^2} \times X_{2,1}$$

$$\vdots$$

$$\hat{X}_{t,1} = \sqrt{1-\phi^2} \times X_{t,1}$$

Then the least squares estimation is done for the entire observations. The Prais-Winsten procedure is limited to the error term with an AR(1) process.

When we employ the Cochrane-Orcutt or Prais-Winsten procedure, we need to utilize the estimated autoregressive process in the residuals, as well as

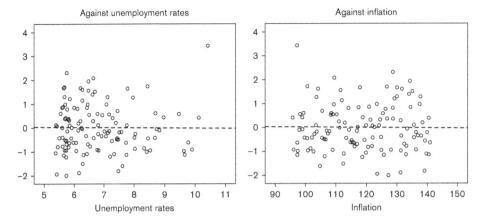

FIGURE 55. Standardized residuals of the multiple time series analysis of monthly violent crime rates, 1983–1992.

the estimated coefficients of the independent variables, to forecast future values of the dependent variable. Since we do not know the future residuals, we need to forecast them one by one (see chapter 4). For example, if the residuals contain an AR(1) process, we forecast the lag-1 future residual, $E_{t+1}$, with the current residual, $\phi E_t$, the lag-2 future residual, $E_{t+2}$, with the forecasted lag-1 future residual, $\phi \hat{E}_{t+1}$, and so on.

*The Case of Monthly Violent Crime Rates, 1983–1992.* For the multiple time series analysis of the monthly violent crime rates (column 1 in table 10), the first three properties of residuals are satisfied. The mean of the residuals is zero. According to the scatter plots of the standardized residuals (figure 55), the residual variance seems to be reasonably constant. According to the stem-and-leaf plot (figure 56) of the standardized residuals, the residual distribution seems to be normal. The normal Q–Q plot also supports the normality of the residual distribution (figure 57).

The autocorrelated-residuals problem was solved after we prewhitened the dependent variable and the two independent variables (figure 51). Including lagged dependent variables did not help us solve the autocorrelated-residuals problem, as Durbin's *h* test statistic is 0.09 (column 1, table 11).[29] When we prewhitened the dependent variable and the two independent variables, the lagged dependent variable became insignificant. The lagged independent variables did not significantly contribute to the explanation of the dependent variable, while they caused very serious multicollinearity problems.

| Stem | Leaf | Frequency |
|------|------|-----------|
| -1. | 558899 | 6 |
| -1. | 0000111122 | 10 |
| -0. | 555666666677788899999999 | 24 |
| -0. | 00001111111223334444444444 | 26 |
| 0. | 0000011122233334444 | 19 |
| 0. | 55555567788888999 | 17 |
| 1. | 122234 | 6 |
| 1. | 556666679 | 9 |
| 2. | 03 | 2 |

Extremes (≥ 3.4)   1
Stem width: 1.00000
Each leaf: 1 case(s)

FIGURE 56. Stem-and-leaf plot of the standardized residuals of the multiple time series analysis of monthly violent crime rates, 1983–1992.

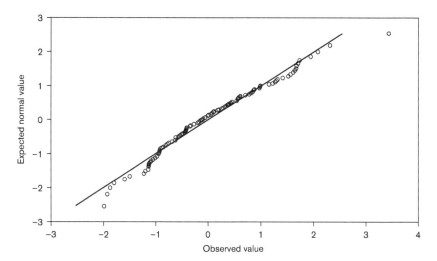

FIGURE 57. Normal Q–Q plot of the residuals from the regression of monthly violent crime rates on consumer price index, 1983–1992.

Instead of prewhitening the dependent variable and the two independent variables to solve the autocorrelated-residuals problem, we can estimate a multiple time series model with AR residuals. According to figures 41 and 42, the residuals of the model with the unwhitened dependent variable and the two unwhitened independent variables (table 9) are likely to contain an AR(1) process with seasonality of lag 12. Before we estimate adjusted regression coefficients and their standard errors with the Cochrane-Orcutt procedure, we deseasonalized the dependent variable to estimate its relationship with the two independent variables more accurately. The estimated coefficient of an AR(1) process is 0.71 (standard error 0.066).[30] We transformed the dependent variable and the two independent variables with the estimated AR coefficient. Then we reran multiple time series analysis with the transformed variables. The adjusted multiple time series model is:[31]

$$\hat{Y}_t = -03.41 + .17 \times \text{unemployment rate} + .471 \times \text{inflation} + 0.71 E_{t-1} + Z_t$$

where $Z_t = WN(0, 2.79)$.

Only inflation appears to significantly influence the monthly violent crime rate (standard error 0.042).[32] About a two-point increase of inflation tended to lead instantaneously to about a one-point increase of the monthly violent crime rate. As expected, the standard error slightly increased, compared with that in the original model (table 9). The standard error appears to be underestimated in the presence of a positive autocorrelation in the residuals in table 9.

In the above equation, $\{Z_t\}$ does not contain any systematic patterns (figures 58 and 59). The Durbin-Watson $d$ statistic changed from 0.736 for $\{E_t\}$ to 1.935 for $\{Z_t\}$. Because the number of independent variables, including the constant, is three, and the number of observations is 120, the lower and upper bounds of the Durbin-Watson $d$ statistic are 1.613 and 1.736, respectively. We failed to reject the null hypothesis that there is no positive first-order autocorrelation in the residuals before we applied the Cochrane-Orcutt procedure. After we adjusted the regression model with the Cochrane-Orcutt procedure, we reject the null hypothesis.

The Breusch-Godfrey statistic changed from 46.846 for $\{E_t\}$ to 0.354 for $\{Z_t\}$.[33] The chi-squared value at $\alpha = .05$ with one degree of freedom is 3.841. Thus, we do not reject the null hypothesis that the coefficient of an AR(1) process is zero for $\{Z_t\}$, but we reject the null hypothesis for $\{E_t\}$.

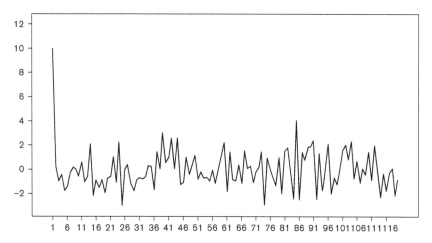

FIGURE 58. Residuals, $\{z_t\}$, of the multiple time series model adjusted with the Cochrane-Orcutt procedure.

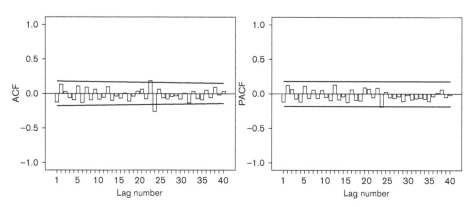

FIGURE 59. ACF and PACF of residuals of the multiple time series model adjusted with the Cochrane-Orcutt procedure.

## (5) FURTHER READING

Achen, C. H. 2000. "Why Lagged Dependent Variables Can Suppress the Explanatory Power of Other Independent Variables." Presented at the Annual Meeting of the Political Methodology Section of the American Political Science Association, Los Angeles, July 20–22.

Beck, N. 1985. "Estimating Dynamic Models is Not Merely a Matter of Technique." *Political Methodology* 11:71–90.

―――. 1993. "The Methodology of Co-integration." *Political Analysis* 4: 237–248.

Beck, N., and J. N. Katz. 2011. "Modeling Dynamics in Time-Series-Cross-Section Political Economy Data." *Annual Review of Political Science* 14:331–352.

Bewley, R. A. 1979. "The Direct Estimation of the Equilibrium Response in a Linear Model." *Economic Letters* 3:357–361.

Box-Steffensmeier, J. M., and R. M. Smith. 1998. "Investigating Political Dynamics Using Fractional Integration Methods." *American Journal of Political Science* 42:661–689.

Box-Steffensmeier, J. M., and A. R. Tomlinson. 2000. "Fractional Integration Methods in Political Science." *Electoral Studies* 19:63–76.

Breusch, T. S. 1978. "Testing for Autocorrelation in Dynamic Linear Models." *Australian Economic Papers* 17:334–355.

Chambers, M. J. 1998. "Long Memory and Aggregation in Macroeconomic Time Series." *International Economic Review* 39:1053–1072.

Cochrane, D., and G. H. Orcutt. 1949. "Application of Least Squares Regression to Relationships Containing Auto-Correlated Error Terms." *Journal of the American Statistical Association* 44:32–61.

De Boef, S., and J. Granato. 1997. "Near-Integrated Data and the Analysis of Political Relationships." *American Journal of Political Science* 41:619–640.

―――. 1999. "Testing for Co-integrating Relationships with Near- Integrated Data." *Political Analysis* 8: 99–117.

De Boef, S., and L. Keele. 2008. "Taking Time Seriously." *American Journal of Political Science* 52:184–200.

Dickey, D. A., and W. A. Fuller. 1979. "Distribution of the Estimators for Autoregressive Time Series with a Unit Root." *Journal of the American Statistical Association* 74:427–431.

Diebold, F. X., and G. D. Rudebusch. 1989. "Long Memory and Persistence in Aggregate Output." *Journal of Monetary Economics* 24:189–209.

Durr, R. 1993. "An Essay on Co-integration and Error Correction Models." *Political Analysis* 4:185–228.

―――. 1993. "Of Forest and Trees." *Political Analysis* 4:255–258.

Engle R. F., and C. W. J. Granger. 1987. "Co-integration and Error Correction: Representation, Estimation and Testing." *Econometrica* 55:251–276.

―――. 1991. *Long-Run Economic Relationships: Readings in Co-integration.* New York: Oxford University Press.

Freeman, J. 1983. "Granger Causality and the Time Series Analysis of Political Relationships." *American Journal of Political Science* 27:327–358.

Freeman, J., D. Houser, P. M. Kellstedt, and J. T. Williams. 1998. "Long-Memoried Processes, Unit Roots, and Causal Inference in Political Science." *American Journal of Political Science* 42:1289–1327.

Godfrey, L. G. 1978. "Testing against General Autoregressive and Moving Average Error Models when the Regressors Include Lagged Dependent Variables." *Econometrica* 46:1293–1302.

Granger, C. W. J. 1969. "Investigating Causal Relations by Econometric Models and Cross-Spectral Methods." *Econometrica* 37:424–438.

———. 1980. "Long Memory Relationships and the Aggregation of Dynamic Models." *Journal of Econometrics* 14:227–238.

Granger, C. W. J., and Z. Ding. 1996. "Varieties of Long Memory Models." *Journal of Econometrics* 73:61–77.

Granger, C. W. J., and R. Joyeux. 1980. "An Introduction to Long-Memory Time Series Models and Fractional Differencing." *Journal of Time Series Analysis* 1:15–29.

Granger, C. W. J., and P. Newbold. 1974. "Spurious Regressions in Econometrics." *Journal of Econometrics* 2:111–120.

Johansen, S. 1988. "Statistical Analysis of Co-integration Vectors." *Journal of Economic Dynamics and Control* 12:231–254.

Keele, L. J., and N. J. Kelly. 2006. "Dynamic Models for Dynamic Theories: The Ins and Outs of Lagged Dependent Variables." *Political Analysis* 14:186–205.

Lebo, M., and H. D. Clarke. 2000. "Modelling Memory and Volatility: Recent Advances in the Analysis of Political Time Series." *Electoral Studies* 19:1–7.

Liu, L. M., and D. H. Hanssens. 1982. "Identification of Multiple-Input Transfer Function Models." *Communications in Statistics—Theory and Methods* 11:297–314.

Maddala, G. S., and I. M. Kim. 2000. *Unit Roots, Co-integration, and Structural Change.* Cambridge, MA: Cambridge University Press.

Murray, M. P. 1994. "A Drunk and Her Dog: An Illustration of Co-integration and Error Correction." *American Statistician* 48:37–39.

Smith, R. 1993. "Error Correction, Attractions, and Co-integration." *Political Analysis* 4:249–254.

Williams, J. T. 1992. "What Goes Around Comes Around: Unit Root Tests and Co-integration." *Political Analysis* 4:229–235.

SEVEN

# Time Series Analysis as an Impact Analysis Method

## (1) INTERRUPTED TIME SERIES ANALYSIS

Impact analysis is used to determine whether an intervention (such as a government program or a law) has the intended impact on the behavior of a target. The impact of an intervention can be identified by assessing the response in a time series to the intervention, namely by looking at the difference between the pre- and post-intervention segments of temporal observations in terms of the segments' levels and/or trends.

Therefore, time series analysis is frequently employed as impact analysis. Time series analysis with an intervention point in the model is termed *interrupted time series analysis* (see e.g. Box and Tiao 1975; Forehand 1982; McDowall et al. 1980). The basic idea of interrupted time series analysis is that if a program has any significant impact, such an impact will be identified by the difference between the observations after the intervention point and the *counterfactual*, the projection of the correctly-modeled-before-the-intervention series into the post-intervention period. The projected series will represent the time series that we would have observed if there had not been the intervention.

The *impact* is a specified consequence of an intervention. There are various patterns of impact. An intervention can cause a change in either or both the level and trend of a time series. Such a change can be temporary or long-lasting. It can be abrupt or gradual. An intervention can have a delayed impact. For example, it may take time for a new crime-control policy to have its intended impact by actually imprisoning criminals. On the other hand, its impact can be observed even before its enactment. An example is a psychologically deterrent impact of a tough crime-control policy on potential criminals. Some possible patterns of impact are depicted in figure 60.

A. Abrupt temporary change in level

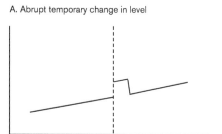

B. Abrupt long-term change in level

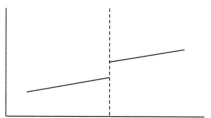

C. Gradual long-term change in level

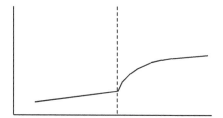

D. Abrupt temporary change in slope or gradual long-term change in level

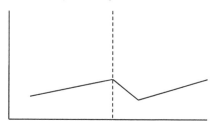

E. Gradual long-term change in slope

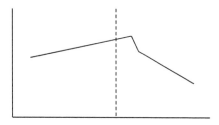

F. Abrupt long-term change in slope

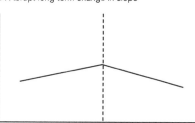

G. Delayed changes in level and slope

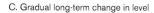

H. Change in slope that precedes the intervention

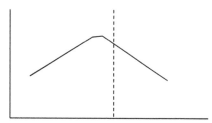

FIGURE 60. Patterns of impact.

As in univariate or multiple time series analysis, discussed in the previous chapters, the first step of interrupted time series analysis is to visually inspect a time series before and after the intervention point. With visual inspection, we can determine whether there is a trend and/or level change after the intervention point, and if so, whether or not the change is in accord with our expectation, whether it is abrupt or gradual, whether it is temporary or long-lasting, whether it is delayed, and whether it precedes the intervention point.

In addition, visual inspection can help us identify seasonality and a systematic noise pattern that should be removed to accurately evaluate the impact of an intervention. When a time series contains these two components, they mask its trend and increase its variance (Makridakis, Wheelwright, and McGee 1983). In interrupted time series analysis, we need to focus on the deterministic trend, not on systematic residual patterns. The impact of an intervention in the social sciences will usually cause changes in the level of a time series and the direction and/or size of the long-term deterministic trend. Changes in systematic residual fluctuations, if any, will not usually convey important theoretical implication with regard to the impact of an intervention.

We can check for the presence of systematic noise patterns and seasonality through visual inspection (see chapter 2). A cyclical periodic pattern in the scatter plots of a time series indicates the presence of seasonality. Smoothness of the scatter plots of a time series, that is, long stretches of plots on the same side above or below a trend line, indicates the presence of systematic noise patterns. When seasonality and systematic noise patterns are present in a time series, we need to deseasonalize and prewhiten the time series for more accurate interrupted time series analysis.

We explained deseasonalization and prewhitening in chapter 2 and chapter 6, section 2. When we estimate and eliminate systematic noise patterns for impact analysis, we need to divide a time series into the before-the-intervention observations and the after-the-intervention observations. When an intervention has a significant impact, the relationship between a time series and Time will significantly change after the intervention. For example, crime rates that have increased may begin decreasing after the enactment of a tough crime-control policy. In this case, if we estimate and eliminate a systematic noise pattern for the entire observations, we may flatten the overall trend. Consequently, the estimated impact may become weaker than it should be.

After we deseasonalize and/or prewhiten a time series, we check for the existence of noticeable change in the observations before and after the intervention, that is whether the after-the-intervention observations show significantly different behavior from the before-the-intervention observations. We can use either univariate time series analysis or multiple time series analysis for this purpose.

In the first method, we simply follow the steps of univariate time series analysis (see chapter 2). We first model the time-dependent behavior of the before-the-intervention observations:

(17) $$Y_t = b_0 + b_1 T + e_t$$

where $T$ represents time scored consecutively from 1 to the last temporal point of the before-the-intervention observations, and $e_t$ is an error term.

We estimate $b_0$ and $b_1$, and then, based on the estimated model, we predict the after-the-intervention series. Since we deseasonalized and prewhitened the before-the-intervention series, $e_t$ should be white noise. The predicted after-the-intervention values serve as a counterfactual: the values we would probably have observed if the intervention had no impact. We expect these predicted values and the after-the-intervention observations to contain significantly different levels and/or slopes.

To estimate the difference in levels and slopes of the counterfactuals and the after-the-intervention observations, we model the time-dependent behavior of the after-the-intervention observations. In this model, $T$ is scored consecutively from 1 (the first observation after the intervention point) to the last observation. To check whether the level changes before and after the intervention, we compare the two values of the dependent variable at which the estimated trend of the before-the-intervention series and that of the after-the-intervention series cross the intervention point. To check whether the slope changes before and after the intervention, we compare the estimated slopes of the before- and after-the-intervention series.

The other method directly incorporates the intervention into the model by additionally including variable(s) that represent the intervention. We need to determine through visual inspection whether we need to model the change in level, slope, or both. First, when the intervention brings about changes only in level, we need to include only an intervention dummy variable that is zero for the before-the-intervention time points and one for the intervention point and the after-the-intervention points:

(18) $$Y_t = b_0 + b_1 T + b_2 I + e_t$$

where $T$ represents time scored consecutively from one to the last temporal point of observations, $I$ is the intervention dummy variable, $b_0$ is the intercept, and $e_t$ is the error term.

Because the intervention dummy variable, $I$, has only two values, $b_2$ describes the change in the overall level before and after the intervention, the vertical difference at the intervention point between the before-the-intervention trend line and the after-the-intervention trend line.

Second, if we observe changes only in slope, we include an additional time variable, $T_2$, not the intervention dummy variable. This additional time variable is zero for the before-the-intervention time points and scores the intervention point and after-the-intervention points consecutively from 1 to the last temporal point:

$$(19) \qquad Y_t = b_0 + b_1 T + b_2 T_2 + e_t$$

In this equation, the slope will be $b_1$ before the intervention, and $b_1 + b_2$ after the intervention.

Third, when we observe a significant change both in level and in slope, we include both $I$ and $T_2$ in the model:

$$(20) \qquad Y_t = b_0 + b_1 T + b_2 T_2 + b_3 I + e_t$$

This model is similar to the dummy variable interactive OLS regression model that we employ to consider the non-additivity of the relationship between two cross-sectional variables. When the relationship between the dependent variable and a continuous independent variable is clearly different for the two subgroups of the independent variable, we include a dummy variable that is zero for the first group and one for the second group and its interaction term with the independent variable. The slope for the first group will be the coefficient of the independent variable, while the slope for the second group will be the sum of the coefficients of the independent variable and the interaction term (Berry and Feldman 1985, 51–72).

In equation (20), $T$ represents time scored from one to the last temporal point of observations; $T_2$ is zero before the intervention, and it begins counting the intervention point and after-the-intervention points consecutively from one to the last temporal point. In this equation, $b_0$ is the overall level of $Y$ and $b_1$ is the trend of $Y$ before intervention. After intervention, the overall level will be $b_0 + b_3$ and the trend will be $b_1 + b_2$.

We may simply try equation (20) without visually inspecting the before- and after-the-intervention observations. If the overall level does not significantly change, $b_3$ will be statistically insignificant. If the slope does not significantly change, $b_2$ will be statistically insignificant. If the intervention does not have any significant impact on level and slope, the level and slope of $Y$ will remain the same ($b_0$ and $b_1$, respectively).

However, it is recommended to visually inspect a time series first. There is a big difference between doing what we know we should do and simply doing something with no definite object in mind and seeing what comes out. By visually inspecting the distribution of observations before and after the intervention, we can determine whether there is a significant change in level or slope or both, and whether the change is temporary. Accordingly, we are ready to estimate and interpret the impact more accurately than when we simply conduct statistical analysis without visually inspecting a time series.

For example, if an intervention brings about a temporary change only in slope, fitting a linear trend to the after-the-intervention observations may produce an inaccurate estimate of the after-the-intervention trend. In figure 61, the intervention changed the slope from positive to negative but the impact did not persist long and the slope returned to its original value. If we simply conduct an interrupted time series analysis without considering this point, the trend estimate for the after-the-intervention observations will give us an inaccurate impression that the intervention has a persistent impact in the level and slope. It will not accurately capture the short-term and long-term dynamics after the intervention.

If the impact of an intervention precedes the intervention point, for example if an intervention has a psychologically deterrent effect, fitting a linear trend to the before-the-intervention observations may produce inaccurate estimates of the before-the-intervention level and slope. If the impact is delayed, fitting a linear trend to the after-the-intervention observations may produce inaccurate estimates of the after-the-intervention level and slope. In figure 62, for example, the amount of drop in the level after the intervention is likely to be underestimated, while the negative coefficient of the after-the-intervention slope is likely to be overestimated.

When nonlinearity of slope exists in either or both the before- and after-the-intervention observations, we may inaccurately estimate the impact. Equations (17)–(20) assume that the relationship between the dependent variable and Time is linear from the first observation to the last. However, after the intervention, the trend may increase or decrease at an obviously

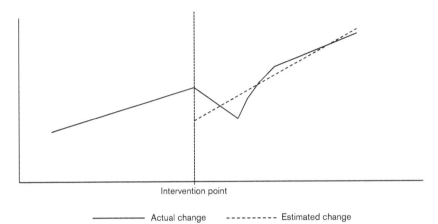

FIGURE 61. An impact that temporarily changed a positive slope into a negative slope.

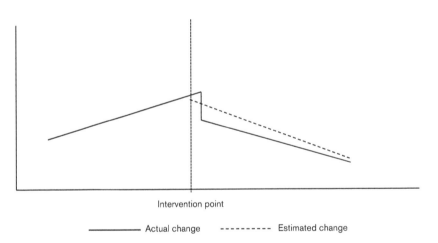

FIGURE 62. A delayed impact on the level.

increasing or decreasing rate. In such a case, we may not accurately estimate the impact of an intervention on the dependent variable. For example, in figure 63A, a curvilinear relationship exists between the dependent variable and Time, and the intervention may not cause a significant change in the level and trend of this time series. However, the OLS estimates of the before- and after-the-intervention slopes may provide an inaccurate impression that the trend changed after the intervention. In Figure 63B, the trend linearly

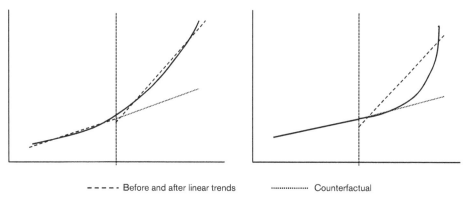

A. Overall curvilinear trend    B. Linear before-trend and curvilinear after-trend

– – – – Before and after linear trends    ·············· Counterfactual

FIGURE 63. Impact analysis when nonlinearity exists in the trend.

increases before the intervention, but it begins increasing at an increasing rate after the intervention. In this case, the estimated after-the-intervention OLS slope may be less accurate, but the difference between the before- and after-the-intervention slopes can accurately tell us that the trend changed after the intervention.

In these cases, we had better employ separate univariate analyses (equation (17)) of the before- or after-the-intervention crime rates, rather than equations (18)–(20). In equations (18)–(20), $T$, $T_2$, and $I$ equally weight all observations, including those which are related to temporary change, psychologically deterrent impact, or delayed impact. Therefore, we may underestimate or overestimate the levels and/or slopes of the before- or after-the-intervention crime rates. In the case of figures 61 and 62, we may need to separately analyze observations related to temporary change, psychologically deterrent impact, or delayed impact. Without these observations, we can more accurately estimate the levels and slopes of the before- or after-the-intervention crime rates. In the case of figures 63A and 63B, we can fit a curvilinear regression model to the before- and/or after-the-intervention observations and provide a supplementary explanation of the characteristic of the changes after the intervention.

Interrupted time series models can be expanded to include more than one intervention point. Most impact analyses include one intervention point, such as the implementation of a specific policy, and the variable that measures the intervention divides a time series into two parts. However, the number of interventions does not have to be limited to one. A government may implement characteristically identical policies for a long period. For example, the UN may repeatedly take sanctions against a country to improve its political situation. In this case, we first divide the time series into sub-groups according to each intervention. Then we include more than one dummy variable ($I$'s) and more than one additional time variables ($T_2$'s) in equation (20). Each $I$ and $T_2$ represents the impact of each intervention on the level and slope of each subgroup. With this expanded interrupted time series model, we can test the impact of one or more interventions on the target behavior (see e.g. Lewis-Beck and Alford 1980). This is identical to conducting separate before- and after-the-intervention time series analyses for each intervention point.

In multiple time series analysis, we often include control variables to measure the coefficients of independent variables more accurately. However, we must pay careful attention to the following possibilities, when we expand interrupted time series models by including factors that are correlated with the dependent variable. First, the estimated coefficients of these control variables will describe the average degree of their covariance with the dependent variable during the entire period of study. If an intervention changes the target behavior of the dependent variable, the estimated coefficients of control variables may convey misleading information about their relationship with the dependent variable. For instance, we may observe a positive coefficient of a control variable during the before-the-intervention period. If we consider only the after-the-intervention period, the coefficient of the control variable may be negative. When we consider both periods, we may observe a positive coefficient of the control variable simply because its covariance with the dependent variable is stronger before the intervention point than after. In this case, the control variable does not help us estimate the impact of the intervention more accurately. It may bias the estimated coefficients of the three main independent variables in equation (20), instead.

Second, if a control variable is significantly correlated with the dependent variable, its relationship with the dependent variable is expected to be reflected in the coefficient of $T$, which depicts the behavior of the dependent variable during the period of the study (see e.g. Kellough 1989). If the control variable had changed in its level and/or trend after the intervention point, for whatever reasons, such changes might have been reflected in the coefficients of $T_2$ and/or $I$, which represent the change in the level and trend, respectively, of the dependent variable after the intervention point. Therefore, including factors that are significantly correlated with the dependent variable in the interrupted time series model amounts to including the same variable twice.

Third, an additional control variable, whether or not it is significantly correlated with the dependent variable, may cause a very serious multicollinearity problem with the two main independent variables, $T$ and $T_2$. These two independent variables increase one unit each time from the first observation of the entire time series or from the intervention point, respectively. Therefore, if the control variable is trending, it is likely to be strongly correlated with these two independent variables. For example, price tends to increase over time. Therefore, when we include inflation in an interrupted time series model, it tends to be highly correlated especially with the time variable $(T)$. The levels of tolerance for both variables often near zero.

To test whether the observed changes in the level and/or slope of the dependent variable were caused by factors that are correlated with the dependent variable, we can separately check whether the levels and/or slopes of these factors change before and after the intervention and whether such changes can explain the changes in the level and/or slope of the dependent variable. In an experiment, we can compare randomly selected two groups, one with an treatment (intervention) and the other without the treatment. Then we do not have to control factors that may influence the dependent variable. However, in the social sciences, it is usually difficult to test the impact of an intervention with an experiment.

With interrupted time series analysis, we test the impact of an intervention that lasts for a considerable period (such as a government program or law). Interrupted time series analysis is not appropriate for the impact of an event that occurs only once. Even if we observe that the behavior of a time series changed in terms of its level and/or slope after the event, we cannot assure ourselves that the event continuously caused the changing behavior. For an event that occurs once, we can check whether the behavior of a time series changes after the event and at which rate the change dissipates per time interval with the estimated autoregressive process of the time series.

For example, a historical event (such as a terrorist attack or a political scandal) may exert an influence only once, and we expect its impact to diminish with the lapse of time. In this case, if we employ interrupted time series analysis that includes the intervention dummy variable, $I$, and additional time variable, $T_2$, we inaccurately assume that an event continued to exist and to exert influence after the time point at which the event occurred. Although the event existed only at one time point, $I$ will assign one to the time point at which the event occurred and to all the after-the-event time points, and $T_2$ will score the time point at which the event occurred and all the after-the-event time points from one to the last temporal point. Therefore, their coefficients will consider all deviations from the mean for the dependent variable in calculating the covariance of $I$ or $T_2$ with the dependent variable.

## (2) THE CASE OF MONTHLY VIOLENT CRIME RATES, 1985–2004

In the early 1990s, violent crime was deemed a continuing social disaster. This led to the enactment of a series of getting-tough-on-crime policies. The federal government enacted the Violent Crime Control and Law Enforcement Act (Public Law 103–322) in September 1994. Its major provisions include:

1. Requiring a life sentence in prison for any individual who is convicted of a third serious violent or drug crime.
2. Increasing the number of crimes subject to the death penalty.

By 1995, 25 state governments had enacted the Three Strikes and You're Out law (TSO, hereafter) or an equivalent law.[1]

Many studies of the impact of the TSO have provided mixed evidence of the getting-tough-on-crime policy's impact (see e.g. Austin, Clark, Hardyman, and Henry 1999; Chen 2008; Ekland-Olson and Kelly 1992; Kovandzic, Sloan III, and Vieraitis 2004; Ramirez and Crano 2003; Schiraldi, Colburn, and Lotke 2004; Stolzenberg and D'Alessio 1997; Worrall 2004). These studies focused on subnational-level statistics, such as crime rates in selected states, counties, or cities. With the getting-tough-on-crime laws being adopted by the federal government and half of the state governments, however, the impact of such laws is expected to be observed nationwide over time. Here, we evaluate the impact of the getting-tough-on-crime policy on national monthly violent crime rates.

U.S. Public Law 102–322 was enacted in September 1994. In addition, getting-tough-on-crime laws had taken effect in 25 states by 1995. We examine the monthly violent crime rates during the 20-year period from January 1985 to December 2004 (10 years up to 1994 and 10 years after 1995). There were 240 monthly observations (117 observations up to September 1994 and 123 observations after October 1994).

The first step in testing the impact of the getting-tough-on-crime policy on the violent crime rate is to visually inspect the violent crime rate, checking whether it changed in the expected direction after the implementation of the policy. Visual inspection will also help us identify and eliminate systematic noise patterns for a more accurate time series analysis.

We assume that the getting-tough-on-crime policy had the intended impact. Therefore, we expect the violent crime rate to have decreased significantly after October 1994. The decrease would be marked as a change in the intercept and/or in the slope of the time series line. The decrease could begin earlier or later than October 1994. For example, the TSO may have had a psychologically deterrent effect. The getting-tough-on-crime policy could be a strong deterrent against criminal activities, making potential offenders refrain from committing crimes. In this case, we would be likely to observe the decrease in crime rates before October 1994. On the other hand, it may have taken time for the TSO to have its intended impact, for example, by actually increasing the number of imprisoned felons and their length of imprisonment. In this case, we would be likely to observe a delayed impact.

Figure 64 reports the monthly violent crime rates from 1985 to 2004. As shown in the figure, the violent crime rates contain seasonal components of lag 12. In addition, the smoothness of the graph suggests the presence of a systematic noise pattern. We first estimated and eliminated the seasonality and the systematic pattern in residuals for a more accurate estimation of the trend. The additive method was used for deseasonalization because the range of seasonal components is not apparently proportional to the level of violent crime rates (see chapter 2).

According to the seasonal components estimated by the additive method (table 13), violent crimes tended to be committed more frequently during warm periods and less frequently during cold periods. For instance, the seasonal component for February tended to pull down the overall crime rates by 7.84 points on average during the 20-year period included in the time series modeling. The seasonal component for August tended to pull up the overall crime rates by 5.42 points on average.

Since the violent crime rates consist of two separate time series with their own linear trends, we prewhitened each of them with their estimated systematic noise pattern. The deseasonalized and detrended crime rates until September 1994 contained an AR(2) process:

$$e_t = -0.471 + 0.466e_{t-1} + 0.383e_{t-2}$$

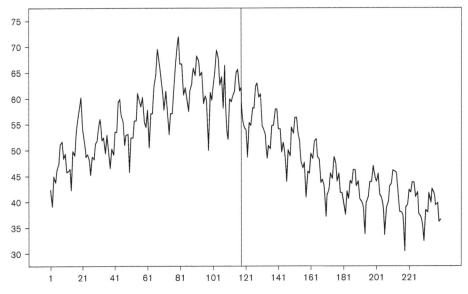

FIGURE 64. Monthly violent crime rates, 1985–2004.

TABLE 13. Estimated Seasonal Components of the
Monthly Violent Crime Rates, 1985–2004

| Month | Seasonal component |
|---|---|
| January | –2.89 |
| February | –7.84 |
| March | –1.98 |
| April | –2.21 |
| May | 1.50 |
| June | 1.78 |
| July | 5.36 |
| August | 5.42 |
| September | 2.16 |
| October | 2.31 |
| November | –2.28 |
| December | –1.33 |

NOTE. The additive method was used to estimate seasonal components.

The deseasonalized and detrended crime rates after October 1994 contained an AR(1) process:

$$e_t = 0.245 + 0.846e_{t-1}$$

We eliminated these autoregressive processes from the deseasonalized violent crime rates. We estimated the impact of the getting-tough-on-crime policy on these deseasonalized and prewhitened crime rates.

According to figure 65, the deseasonalized and prewhitened crime rates show a clear upward trend until December 1993 (the 108th point). It is not clear when the downward trend began. The downward trend became obvious as of January 1995 (the 121st point). However, the downward trend might be traced to January 1994 (the 109th point).

Public Law 102–322 was proposed October 26, 1993, and enacted September 13, 1994. Thus, violent crime rates began decreasing soon after the federal government initiated its getting-tough-on-crime approach. This change before October 1994 may demonstrate the psychologically deterrent effect of the federal getting-tough-on-crime policy. Otherwise, this change may be attributable to some states that began implementing the TSO or an equivalent law earlier. The TSO or an equivalent law was enacted by states after December 1993. In Washington, the first state that enacted the TSO, voters approved Initiative 593—a ballot measure dictating that three-time serious offenders be sentenced to life in prison without parole—in November 1993. Initiative 593 became effective December 2, 1993. Twelve states enacted the TSO or an equivalent law in 1994.

To estimate the impact of the getting-tough-on-crime policy on the violent crime rate, we employ the following interrupted time series model with October 1994 as the intervention point:

$$Y_t = b_0 + b_1 T + b_2 I + b_3 T_2 + e_t$$

where $Y_t$ is the monthly violent crime rates; $T$ is a time variable, which ranges from 1 (January 1985) to 240 (December 2004); $I$ is a dummy variable scored 0 for observations until September 1994 and 1 for observations from October 1994; $T_2$ is an interaction term between $T$ and $I$, scored 0 for observations until September 1994 and 1 (October 1994), 2 (November 1994), ... , 123 (December 2004); and $e_t$ is an error term.

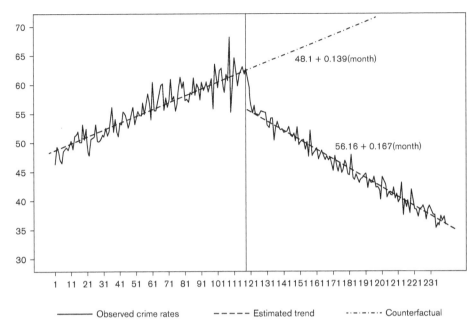

| 70 |
| 65 | 48.1 + 0.139(month) |
| 60 |
| 55 |
| 50 | 56.16 + 0.167(month) |
| 45 |
| 40 |
| 35 |
| 30 |

1  11  21  31  41  51  61  71  81  91  101 111 121 131 141 151 161 171 181 191 201 211 221 231

——— Observed crime rates    – – – – Estimated trend    ·· – · – ·· Counterfactual

FIGURE 65. Deseasonalized and prewhitened monthly violent crime rates, 1985–2004.

In the equation, $b_1$ represents the trend in crime rates before the intervention, while $b_2$ and $b_3$ tell us the change of the level and trend, respectively, in violent crime rates after October 1994. If the getting-tough-on-crime policy had significant impact on crime reduction, $b_2$ and $b_3$ would significantly differ from zero and have negative values.[2]

Table 14 reports the results of the interrupted time series analysis. The national violent crime rates clearly leveled off and began declining after October 1994. The three independent variables ($T$, $I$, and $T_2$) together clearly show that the getting-tough-on-crime policy had its intended impact. According to $b_2$, the level of the violent crime rates decreased by about 9.2 points after the intervention point. In addition, according to $b_1$ and $b_3$, the trend clearly changed its direction: it tended to increase (slope = 0.139) before the intervention point but to decrease ($-0.305 + 0.139 = -0.166$) afterwards. Consequently, the violent crime rate actually declined after the intervention point by about 25.48 points (41.04%), from 62.08 (October 1994) to 36.6 (December 2004) crimes per 100,000 inhabitants.

The intercept in regression analysis is the value of the dependent variable when the independent variable(s) are zero. In figure 65, the intercept for the

TABLE 14. Interrupted Time Series Analysis of the Impact of the Getting-Tough-on-Crime Policy on Monthly Violent Crime Rates, 1985–2004

| | Coefficient | Standard error |
|---|---|---|
| $X_{1t}$ | .139*** | .005 |
| $X_{2t}$ | −9.162*** | .462 |
| $X_{3t}$ | −.305*** | .007 |
| Intercept | 48.101*** | .329 |
| $N$ | 240 | |
| Adjusted $R^2$ | .948 | |
| Durbin-Watson $d$ | 1.655 | |
| $P$ | < .000 | |

***$p$ < .001 (two-tailed).

after-the-intervention crime rates, 56.162, was the intercept of the univariate time series model of the after-the-intervention crime rates. It is located at the last month of the before-the-intervention period (September 1994, $X_1$ = 117). The crime rate for the 117th point that was predicted with the univariate time series model of the before-the-intervention crime rates was 65.3 crimes per 100,000 inhabitants. Thus, the drop before and after the intervention is about 9.14 points, which is close to the estimated coefficient of $I$ (9.2).[3] According to the estimated interrupted time series model, the predicted crime rate is 64.364 for September 1994 ($X_1$ = 117, $X_2$ = 0, and $X_3$ = 0) and 55.036 for October 1994 ($X_1$ = 118, $X_2$ = 1, and $X_3$ = 1). Thus, the level fell by about 9.3 points one month after the intervention. The size of decrease is slightly greater than 9.14 because the negative after-the-intervention slope is reflected.

The violent crime rates after the intervention point were clearly different from the counterfactual, the projection of the modeled-before-the-intervention series into the post-intervention period—that is, the violent crime rates we would have observed if there had been no intervention (figure 65). The difference between the crime rate predicted by the before-the-intervention model and the observed crime rate in December 2004 was 47.273.[4]

It may be controversial to treat the trend from January 1994 to September 1994 as representing the psychologically deterrent effect of the TSO. Some may argue that the downward trend simply began before the federal and state governments implemented the TSO. In this case, the downward trend may serve as evidence that the decrease in violent crime rates is not particularly attributable to the TSO. To check this point, we estimated the trend of violent crime rates only during this period, although the trend estimate may not be reliable because of the small number of cases (9). The trend during this period is not significant (0.50, standard error 0.306). In addition, the intercept (60.7, standard error 1.72) is similar to the crime rate (63.11) that was predicted by the before-the-intervention-series model at that point (the 109th point). Thus, considering this result together with table 14, we can conclude:

1. The national violent crime rate slightly leveled off and stopped increasing as the getting-tough-on-crime approach was initiated in 1993.
2. The national violent crime rate clearly leveled off and began decreasing as the federal government (and many state governments) enacted the getting-tough-on-crime laws in 1994.


## (3) FURTHER READING

Box, G. E. P., and G. C. Tiao. 1975. "Intervention Analysis with Applications to Economic and Environmental Problems." *Journal of the American Statistical Association* 70:70–79.

Cook, T. D., and D. T. Campbell. 1979. *Quasi-Experimentation: Design & Analysis Issues for Field Settings.* Boston, MA: Houghton Mifflin.

Corporale, T., and K Greier. 2005. "How Smart Is My Dummy? Time Series Tests for the Influence of Politics." *Political Analysis* 13:77–94.

Forehand, G. 1982. *New Directions for Program Evaluation: Applications of Time Series Analysis to Evaluation.* San Francisco, CA: Jossey-Bass.

McDowall, D., R. McCleary, E. E. Meidinger, and R. A. Hay Jr. 1980. *Interrupted Time Series Analysis.* Sage University Paper Series on Quantitative Applications in the Social Sciences, no. 21. Newbury Park, CA: Sage.

Mohr, L. B. 1992. *Impact Analysis for Program Evaluation.* Newbury Park, CA: Sage.

Rossi, P. H., and H. E. Freeman. 1993. *Evaluation: A Systematic Approach.* 5th ed. Newbury Park, CA: Sage.

# Links to Online Time Series Analysis Program Manuals

The following links bring you directly to each program's online manual for time series analysis.

- EViews
  http://www.eviews.com/illustrated/illustrated.html
  http://www.eviews.com/Learning/index.html

- Matlabs
  http://www.mathworks.com/help/econ/time-series-regression-models.html
  http://www.mathworks.com/help/econ/arima-class.html
  http://www.mathworks.com/help/econ/regarima-class.html

- Minitab
  http://support.minitab.com/en-us/minitab/17/topic-library/
  modeling-statistics/time-series/basics/methods-for-analyzing-time-series/

- R
  https://cran.r-project.org/web/views/TimeSeries.html
  http://a-little-book-of-r-for-time-series.readthedocs.org/en/latest/index.html
  R is free software, and there are many websites that provide a detailed explanation about how to use R for time series analysis. For example, see arima.html
  and others in the folder at http://search.r-project.org/R/library/stats/html and
  reference manuals for the forecast package, the timsac package, and others at
  https://cran.r-project.org/web/packages/available_packages_by_name.html.

- SAS
  http://support.sas.com/documentation/cdl/en/etsug/68148/HTML/default
  /viewer.htm

http://support.sas.com/documentation/cdl/en/etsug/68148/PDF/default/etsug.pdf

http://support.sas.com/rnd/app/ets/#s1=2

- S+

  http://www.solutionmetrics.com.au/support/splus82win/statman2.pdf

- SPSS

  ftp://public.dhe.ibm.com/software/analytics/spss/documentation/statistics/24.0/en/client/Manuals/IBM_SPSS_Forecasting.pdf

- Stata

  http://www.stata.com/bookstore/time-series-reference-manual/

- Statgraphics

  http://cdn2.hubspot.net/hubfs/402067/PDFs/How_To_Forecast_Seasonal_Time_Series_Data.pdf

  http://cdn2.hubspot.net/hubfs/402067/PDFs/How_To_Construct_a_Control_Chart_for_Autocorrelated_Data.pdf

# US Monthly Violent Crime Rates per 100,000 Inhabitants, 1983–2004

| | 1983 | 1984 | 1985 | 1986 | 1987 | 1988 | 1989 | 1990 | 1991 | 1992 | 1993 |
|---|---|---|---|---|---|---|---|---|---|---|---|
| January | 45.23 | 41.78 | 42.41 | 46.46 | 48.17 | 49.06 | 53.05 | 57.81 | 57.62 | 60.00 | 59.69 |
| February | 40.32 | 40.27 | 39.07 | 42.27 | 45.12 | 46.52 | 45.75 | 50.49 | 53.07 | 57.57 | 49.99 |
| March | 43.29 | 42.63 | 44.95 | 49.85 | 48.78 | 50.34 | 52.38 | 57.08 | 57.08 | 61.36 | 61.18 |
| April | 42.39 | 41.91 | 43.70 | 48.90 | 48.17 | 49.06 | 52.38 | 57.08 | 57.08 | 62.87 | 59.69 |
| May | 44.42 | 44.56 | 46.16 | 52.88 | 51.21 | 53.52 | 55.70 | 62.20 | 62.93 | 65.90 | 62.67 |
| June | 44.47 | 45.62 | 47.25 | 55.78 | 51.82 | 53.52 | 55.70 | 64.40 | 65.95 | 64.39 | 64.91 |
| July | 48.54 | 48.30 | 51.14 | 58.04 | 54.87 | 59.26 | 61.0 | 69.52 | 69.75 | 68.18 | 69.39 |
| August | 49.94 | 49.37 | 51.68 | 60.22 | 56.09 | 59.90 | 59.68 | 66.59 | 72.02 | 67.42 | 67.90 |
| September | 46.83 | 46.32 | 48.33 | 54.05 | 51.82 | 56.71 | 58.35 | 64.40 | 66.71 | 64.39 | 62.67 |
| October | 46.41 | 48.23 | 49.40 | 52.20 | 52.43 | 55.44 | 60.34 | 62.20 | 66.71 | 65.15 | 64.16 |
| November | 42.95 | 42.97 | 45.75 | 48.63 | 49.39 | 50.98 | 55.70 | 57.81 | 60.65 | 59.09 | 58.20 |
| December | 43.44 | 47.23 | 45.90 | 49.24 | 53.04 | 52.89 | 54.37 | 61.47 | 62.16 | 60.60 | 66.40 |

| | 1994 | 1995 | 1996 | 1997 | 1998 | 1999 | 2000 | 2001 | 2002 | 2003 | 2004 |
|---|---|---|---|---|---|---|---|---|---|---|---|
| January | 54.95 | 54.08 | 52.84 | 49.49 | 47.68 | 42.89 | 39.51 | 38.80 | 39.06 | 37.10 | 36.10 |
| February | 52.09 | 48.60 | 48.38 | 43.99 | 40.87 | 37.13 | 37.48 | 33.80 | 33.60 | 30.45 | 32.40 |
| March | 59.94 | 55.44 | 50.93 | 50.10 | 45.98 | 41.32 | 42.04 | 39.90 | 39.10 | 39.02 | 38.40 |
| April | 59.23 | 54.76 | 50.29 | 48.88 | 45.41 | 42.36 | 40.52 | 40.90 | 40.00 | 39.49 | 38.00 |
| May | 60.66 | 58.18 | 54.75 | 54.38 | 49.38 | 45.50 | 44.07 | 43.90 | 43.01 | 42.30 | 41.69 |
| June | 61.37 | 58.18 | 54.75 | 53.16 | 48.25 | 44.46 | 43.56 | 43.90 | 43.50 | 41.87 | 39.84 |
| July | 64.94 | 62.29 | 57.93 | 56.21 | 51.65 | 48.64 | 46.09 | 46.90 | 46.00 | 43.77 | 42.60 |
| August | 65.65 | 62.97 | 57.93 | 56.21 | 52.22 | 47.59 | 46.09 | 44.90 | 45.98 | 43.77 | 41.69 |
| September | 61.37 | 60.24 | 54.11 | 53.16 | 48.81 | 43.93 | 43.05 | 43.90 | 45.50 | 40.90 | 39.40 |
| October | 62.08 | 60.92 | 54.11 | 51.94 | 48.25 | 45.50 | 44.07 | 45.40 | 42.50 | 41.90 | 39.84 |
| November | 55.66 | 54.76 | 49.65 | 47.66 | 43.71 | 41.84 | 40.52 | 41.40 | 38.10 | 37.59 | 36.13 |
| December | 54.23 | 54.08 | 51.56 | 46.44 | 44.27 | 41.84 | 40.01 | 40.90 | 38.10 | 37.10 | 36.60 |

SOURCE: *Uniform Crime Reports: Crime in the United States* (Washington, DC: Department of Justice, Federal Bureau of Investigation).

# Data Resources for Social Science Series Analysis

The following are webpages/data resources from which readers can download datasets for their social science time series analysis. This list includes webpages/data resources that provide cross-sectional data sets that have accumulated for a period long enough to be used for time series analysis.

## CRIME

- **Disaster Center Crime Pages** provide data on national and state-level crime rates in the U.S. from 1960 (http://www.disastercenter.com/crime/).
- **Federal Bureau of Investigation's Crime Statistics** provide data on violent crimes, property crimes, hate crimes, and other data on sworn officers and civilian employees in the U.S. (https://www.fbi.gov/stats-services/crimestats, https://www.fbi.gov/about-us/cjis/ucr).
- **National Archive of Criminal Justice Data** provides data on criminological and criminal justice and links to related websites (http://www.icpsr.umich.edu /icpsrweb/NACJD/index.jsp).
- **RAND State Statistics** contain social science databases on multiple topics, such as crimes, demographics, and economics, covering all 50 U.S. states (http://www.randstatestats.org).
- **Transactional Records Access Clearinghouse** provides data on spending and enforcement activities of the federal government: FBI, INS, DEA, Homeland Security, and IRS (http://tracfed.syr.edu/).
- **UK CrimeStats** provides monthly crime rates in the U.K. from 2010 (http:// www.ukcrimestats.com/National_Picture/).

- **American Community Survey** is conducted by the U.S. Census Bureau to supplement the census data that is collected every tenth year. The survey collects data on an ongoing basis, January through December, to provide more current demographic, economic, and housing characteristics at geographic levels comparable to the decennial census (http://www.census.gov/programs-surveys /acs). Tables and maps can be generated from the **American Factfinder** site (http://factfinder.census.gov).

- **British Household Panel Survey** (BHPS) is an annual survey that measures social and economic characteristics at the individual and household level in Britain from 1991–2009 (waves 1–18). From wave 19, the BHPS became part of the **United Kingdom Household Longitudinal Study** (UKHLS). Users can match the BHPS wave 1–18 data to the UKHLS wave 2 data and onwards (https://discover.ukdataservice.ac.uk/series/?sn=200005).

- **Canadian Socioeconomic Information Management database** contains more than 52 million numeric time series covering a wide variety of social and economic indicators for Canada (http://www.statcan.gc.ca/start-debut-eng. html).

- **County and City Data Books** provides access to the 1944–2000 County and City Data Books, which include level of income, level of education, and unemployment rate (http://ccdb.lib.virginia.edu). **QuickFacts** at the U.S. Census Bureau site has up-to-date data for states, counties, and cities (http:// www.census.gov/quickfacts/table/PST045215/00).

- **EUROSTAT** provides data on demography, economy, and finance at the European level (http://ec.europa.eu/eurostat/web/main/home).

- **German Socio-Economic Panel** is a yearly longitudinal study of private households, starting in 1984. The data provide information on all household members, including demography, employment, income, taxes, social security, education, attitudes, values, and personality (https://data.soep.de/topics).

- **Minority Data Resource Center** provides data on issues (crime, education, employment, discrimination, immigration, etc.) affecting racial and ethnic minority populations in the U.S. (http://www.icpsr.umich.edu/icpsrweb/RCMD/).

- **Murray Research Archive** provides data on human development and social change, especially data that illuminate women's lives and issues of concern to women (http://murray.harvard.edu/dataverse).

- **National Archives and Records Administration** provides access to U.S. federal government records related to agriculture, economic and financial statistics, demographics, labor, education, health and social services, attitudinal data, international issues, military, and the environment. Most of the records start in the 1960s, with some from as early as World War II (http://www .archives.gov/research/electronic-records/).

- **National Center for Education Statistics** provides data related to the condition of education in the U.S. and internationally (http://nces.ed.gov).
- **PolicyMap** provides data related to demographics, socioeconomics, mortgages and home sales, health statistics, jobs and employment, and so on, for geographies across the U.S., often at neighborhood scale, such as by census tract and block group (http://www.policymap.com/data.html).
- **U.S. Census Bureau** provides census data regarding population and government (http://www.census.gov/en.html).

ECONOMY

- **Bureau of Economic Analysis** provides information on national economic growth and regional economic development, such as gross domestic product, personal income, consumer spending, and corporate profits, in the U.S.; and international economy, such as trade in goods and services, balance of payments, international investment, and multinational enterprises—some data going back to the 1940s (http://www.bea.gov).
- **Bureau of Labor Statistics** provides time series data on employment, prices, working conditions, and productivity in the U.S. and other countries (http://www.bls.gov/data/home.htm).
- **China Data Online** (Zhongguo shu ju zai xian) includes monthly and yearly reports on China's macroeconomic development, population, and economy at the county and city level, and statistical yearbooks (http://chinadataonline.org).
- **Consumer Expenditure Survey** collects data on the buying habits and expenditures of households in the U.S. by a quarterly interview survey and weekly diary survey from 1989 to 2014. Aggregate expenditure shares tables give total expenditures by category for all consumer units and percentages of the totals for various demographic groups (http://www.bls.gov/cex/csxashar.htm).
- **Consumer Price Index** is a measure of the average change in prices paid by urban consumers for goods and services in the U.S. (http://www.bls.gov/cpi).
- **Federal Reserve Archive** (FRASER) provides economic data such as U.S. federal budget and consumer price index (https://fraser.stlouisfed.org).
- **Federal Reserve Board** provides current and historical data associated with surveys such as the Survey of Consumer Finances and Survey of Small Business Finances (http://www.federalreserve.gov/econresdata/).
- **Inter-American Development Bank** provides data on economic performance and institutions in Latin America from 1960 by topic and by country (https://data.iadb.org/DataCatalog/Dataset).

- **LABORSTA** provides data on economic indicators such as employment, working conditions, income, economic stability, and trade unions for about 200 counties, with some series beginning in 1970 (http://laborsta.ilo.org).
- **National Longitudinal Surveys** collects data on labor market experiences of individuals in the U.S. over time (http://www.nlsinfo.org/dbgator/index.php3 and http://www.bls.gov/nls).
- **Surveys of Consumers** provides monthly, quarterly, and yearly indices of consumer sentiment, current economic conditions, and consumer expectations in the U.S. from 1946 onwards (http://www.sca.isr.umich.edu/tables.html).
- **Tax Foundation** provides data on tax rates, collections and overall tax burdens, comparisons among states, results of opinion polls on taxes, and so on, in the U.S. (http://taxfoundation.org/data).
- **Union Membership and Coverage Database** provides private- and public-sector labor union membership, coverage, and density estimates in the U.S. compiled from the monthly household Current Population Survey. Economy-wide estimates are provided beginning in 1973; estimates by state, detailed industry, and detailed occupation begin in 1983; and estimates by metropolitan area begin in 1986 (http://unionstats.gsu.edu).
- **World Income Inequality Database** provides information on income inequality in developed, developing, and transitional countries (https://www.wider.unu.edu/data).

ELECTIONS

- **American National Election Studies** collects election-related data every two years from 1948 onwards (http://electionstudies.org). It is mostly cross-sectional data and occasionally panel data. The same items (such as party identification, feeling thermometers, candidate evaluations, voting, and vote choices) are collected in every survey, and therefore we can build aggregate-level time series data. In addition, the ANES have merged all cross-section cases and variables for select questions from the ANES conducted since 1948 into a single file, the ANES Time Series Cumulative Data File (http://electionstudies.org/studypages/anes_timeseries_cdf/anes_timeseries_cdf.htm).
- **Atlas of U.S. Presidential Elections** aggregates official election results from all 50 states and D.C. (http://uselectionatlas.org).
- **Canadian Election Study** provides data on election participation and vote choices in Canadian since 1965 (http://www.queensu.ca/cora/ces.html).
- **Cooperative Congressional Election Study** is a 50,000+ person national stratified sample survey administered by YouGov/Polimetrix in the U.S. every year from 2005 to 2012. The survey consists of two waves in election years. In

non-election years, the survey consists of a single wave conducted in the early fall (http://projects.iq.harvard.edu/cces).

- **Election Assistance Commission** provides U.S. national and state-level voter registration and turnout statistics for federal elections from 1960 (http://www.eac.gov/research/election_administration_and_voting_survey.aspx).

- **Federal Election Commission** provides data related to campaign finance (http://www.fec.gov/about.shtml). An FTP site for FEC data files back to 1978 is available (ftp://ftp.fec.gov/FEC).

INTERNATIONAL/COMPARATIVE

- **Council of European Social Science Data Archives** provides access to important resources of relevance to the European social science research agenda. It also provides links to the data centers of member countries: Austria, Belgium, Czech Republic, Denmark, Finland, France, Germany, Greece, Lithuania, Netherlands, Norway, Slovenia, Sweden, Switzerland, and United Kingdom (http://cessda.net/).

- **Cross-National Time-Series Data Archive** is a dataset of annual data for social science research from 1815 onwards for over 200 countries. It includes data related to domestic conflict usage, economy, elections, legislative process, government revenue and expenditure, military, population, and education (http://www.databanksinternational.com).

- **Global Peace Index** is a measure of the relative position of nations' and regions' peacefulness in terms of the level of safety and security in society, the extent of domestic and international conflict, and the degree of militarization since 2007 (http://economicsandpeace.org).

- **Kansas Event Data System** uses automated coding of English-language news reports to generate political event (conflicts and mediation) data focusing on the Middle East, Balkans, and West Africa (https://dataverse.harvard.edu/dataset .xhtml?persistentId=hdl:1902.1/10713).

- **Minorities at Risk Project** monitors the status and conflicts of 284 politically active ethnic groups in all countries with a current population of at least 500,000 throughout the world from 1945 to the present (http://www.cidcm .umd.edu/mar).

- **OECD DATA** provides data on economy, education, finance, government, health, society, and so on for countries in the Organisation for Economic Co-operation and Development (https://data.oecd.org/).

- **UNdata** pools major UN databases and several international databases (such as LABORSTA and OECD DATA), so users do not have to move from one

database to another to access different types of information (http://data
.un.org/).

- **UNESCO Institute for Statistics** provides data on education, science and
technology, culture, and communication for more than 200 countries (http://
www.uis.unesco.org).

- **UN Population Information Network** provides a list of links to national and
nongovernmental agencies that provide data on demographics, population, and
family planning (http://www.un.org/popin/other4.htm).

- **Vote World** archives datasets of roll-call voting from legislative bodies throughout the international community, including the U.S. House of Representatives
and Senate, the European Parliament, and the United Nations (http://
voteworld.berkeley.edu).

## POLITICAL BEHAVIOR

- **American Presidency Project** provides archives that contain datasets on
presidents' relationship with congress, popularity, and so on. It also provides
document archives related to the study of the presidency, from which
researchers can build their own time series data sets, for example regarding the
presidents' issue priority and the degree of going public (http://www.presidency
.ucsb.edu).

- **Center for Responsive Politics** (OpenSecrets.org) offers data related to money
in U.S. politics, such as campaign expenditures and contributions (https://
www.opensecrets.org).

- **Keith Poole's Data Download Page** (http://voteview.com/dwnl.html) and
Tim Groseclose's Interest Group Score Page (http://www.sscnet.ucla.edu/
polisci/faculty/groseclose/Adj.Int.Group.Scores) provide measures of U.S.
legislators' ideological predispositions. These two measures are substitutes
for interest groups' ratings of individual legislators' voting records (such as
American Conservative Union and Americans for Democratic Action ratings)
that are not comparable over time.

- **Pew Research Center** makes its data related to various topics (such as U.S.
Politics & Policy, Journalism & Media, Religion & Public Life, Global
Attitudes & Trends, and Social & Demographic Trends) available to the public
for secondary analysis after a period of time (http://www.pewresearch.org/data
/download-datasets).

- **Policy Agendas Project** collects and organizes data (such as Gallup's "most
important problem," policy moods, and federal budget) from various archived
sources to trace changes in the national policy agenda and public policy
outcomes since the Second World War (http://www.policyagendas.org).

- **Presidential Data Archive (Presidency Research)** provides datasets that include presidential support scores and survey reports from Gallup, CBS/New York Times, Tyndal, Wirthlin, and Pew (http://presdata.tamu.edu).

PUBLIC OPINION

Public opinion surveys provided by the following organizations repeatedly contain the same items, and therefore we can build time series data regarding various topics.

- **Canadian Opinion Research Archive** collects opinion surveys dating back to the 1970s (http://www.queensu.ca/cora).
- **Gallup** provides surveys that track public opinions on various political, economic, and social issues in more than 160 countries (http://www.gallup .com).
- **Inter-university Consortium for Political and Social Research** provides mostly cross-sectional survey data and occasionally panel data about various topics in the U.S. and abroad (http://www.icpsr.umich.edu/icpsrweb/ICPSR).
- **Roper Center for Public Opinion Research** provides social science data, specializing in data from public opinion surveys on a vast range of topics in U.S. and abroad, from the 1930s (http://ropercenter.cornell.edu/).

# Statistical Tables

The following tables are for the Durbin-Watson $d$ test and the Dickey-Fuller unit root $t$ test. Tables for other test statistics, such as percentiles of the chi-squared distribution, percentiles of the $F$, percentiles of the $t$ distribution, and areas under the standardized normal distribution, are widely available, and therefore they are not provided here.

TABLE A1. Durbin-Watson Statistic: 5% Significance Points of $d_L$ and $d_U$ for Models with an Intercept

| | k=1 | | k=2 | | k=3 | | k=4 | | k=5 | | k=6 | | k=7 | | k=8 | | k=9 | | k=10 | |
|---|---|---|---|---|---|---|---|---|---|---|---|---|---|---|---|---|---|---|---|---|
| n | $d_L$ | $d_U$ | $d_L$ | $d_U$ | $d_L$ | $d_U$ | $d_L$ | $d_U$ | $d_L$ | $d_U$ | $d_L$ | $d_U$ | $d_L$ | $d_U$ | $d_L$ | $d_U$ | $d_L$ | $d_U$ | $d_L$ | $d_U$ |
| 6 | 0.610 | 1.400 | | | | | | | | | | | | | | | | | | |
| 7 | 0.700 | 1.356 | 0.467 | 1.896 | | | | | | | | | | | | | | | | |
| 8 | 0.763 | 1.332 | 0.559 | 1.777 | 0.367 | 2.287 | | | | | | | | | | | | | | |
| 9 | 0.824 | 1.320 | 0.629 | 1.699 | 0.455 | 2.128 | 0.296 | 2.588 | | | | | | | | | | | | |
| 10 | 0.879 | 1.320 | 0.697 | 1.641 | 0.525 | 2.016 | 0.376 | 2.414 | 0.243 | 2.822 | | | | | | | | | | |
| 11 | 0.927 | 1.324 | 0.758 | 1.604 | 0.595 | 1.928 | 0.444 | 2.283 | 0.315 | 2.645 | 0.203 | 3.004 | | | | | | | | |
| 12 | 0.971 | 1.331 | 0.812 | 1.579 | 0.658 | 1.864 | 0.512 | 2.177 | 0.380 | 2.506 | 0.268 | 2.832 | 0.171 | 3.149 | | | | | | |
| 13 | 1.010 | 1.340 | 0.861 | 1.562 | 0.715 | 1.816 | 0.574 | 2.094 | 0.444 | 2.390 | 0.328 | 2.692 | 0.230 | 2.985 | 0.147 | 3.266 | | | | |
| 14 | 1.045 | 1.350 | 0.905 | 1.551 | 0.767 | 1.779 | 0.632 | 2.030 | 0.505 | 2.296 | 0.389 | 2.572 | 0.286 | 2.848 | 0.200 | 3.111 | 0.127 | 3.360 | | |
| 15 | 1.077 | 1.361 | 0.946 | 1.543 | 0.814 | 1.750 | 0.685 | 1.977 | 0.562 | 2.220 | 0.447 | 2.471 | 0.343 | 2.727 | 0.251 | 2.979 | 0.175 | 3.216 | 0.111 | 3.438 |
| 16 | 1.106 | 1.371 | 0.982 | 1.539 | 0.857 | 1.728 | 0.734 | 1.935 | 0.615 | 2.157 | 0.502 | 2.388 | 0.398 | 2.624 | 0.304 | 2.860 | 0.222 | 3.090 | 0.155 | 3.304 |
| 17 | 1.133 | 1.381 | 1.015 | 1.536 | 0.897 | 1.710 | 0.779 | 1.900 | 0.664 | 2.104 | 0.554 | 2.318 | 0.451 | 2.537 | 0.356 | 2.757 | 0.272 | 2.975 | 0.198 | 3.184 |
| 18 | 1.158 | 1.391 | 1.046 | 1.535 | 0.933 | 1.696 | 0.820 | 1.872 | 0.710 | 2.060 | 0.603 | 2.258 | 0.502 | 2.461 | 0.407 | 2.668 | 0.321 | 2.873 | 0.244 | 3.073 |
| 19 | 1.180 | 1.401 | 1.074 | 1.536 | 0.967 | 1.685 | 0.859 | 1.848 | 0.752 | 2.023 | 0.649 | 2.206 | 0.549 | 2.396 | 0.456 | 2.589 | 0.369 | 2.783 | 0.290 | 2.974 |
| 20 | 1.201 | 1.411 | 1.100 | 1.537 | 0.998 | 1.676 | 0.894 | 1.828 | 0.792 | 1.991 | 0.691 | 2.162 | 0.595 | 2.339 | 0.502 | 2.521 | 0.416 | 2.704 | 0.336 | 2.885 |
| 21 | 1.221 | 1.420 | 1.125 | 1.538 | 1.026 | 1.669 | 0.927 | 1.812 | 0.829 | 1.964 | 0.731 | 2.124 | 0.637 | 2.290 | 0.546 | 2.461 | 0.461 | 2.633 | 0.380 | 2.806 |
| 22 | 1.239 | 1.429 | 1.147 | 1.541 | 1.053 | 1.664 | 0.958 | 1.797 | 0.863 | 1.940 | 0.769 | 2.090 | 0.677 | 2.246 | 0.588 | 2.407 | 0.504 | 2.571 | 0.424 | 2.735 |
| 23 | 1.257 | 1.437 | 1.168 | 1.543 | 1.078 | 1.660 | 0.986 | 1.785 | 0.895 | 1.920 | 0.804 | 2.061 | 0.715 | 2.208 | 0.628 | 2.360 | 0.545 | 2.514 | 0.465 | 2.670 |
| 24 | 1.273 | 1.446 | 1.188 | 1.546 | 1.101 | 1.656 | 1.013 | 1.775 | 0.925 | 1.902 | 0.837 | 2.035 | 0.750 | 2.174 | 0.666 | 2.318 | 0.584 | 2.464 | 0.506 | 2.613 |
| 25 | 1.288 | 1.454 | 1.206 | 1.550 | 1.123 | 1.654 | 1.038 | 1.767 | 0.953 | 1.886 | 0.868 | 2.013 | 0.784 | 2.144 | 0.702 | 2.280 | 0.621 | 2.419 | 0.544 | 2.560 |
| 26 | 1.302 | 1.461 | 1.224 | 1.553 | 1.143 | 1.652 | 1.062 | 1.759 | 0.979 | 1.873 | 0.897 | 1.992 | 0.816 | 2.117 | 0.735 | 2.246 | 0.657 | 2.379 | 0.581 | 2.513 |
| 27 | 1.316 | 1.469 | 1.240 | 1.556 | 1.162 | 1.651 | 1.084 | 1.753 | 1.004 | 1.861 | 0.925 | 1.974 | 0.845 | 2.093 | 0.767 | 2.216 | 0.691 | 2.342 | 0.616 | 2.470 |
| 28 | 1.328 | 1.476 | 1.255 | 1.560 | 1.181 | 1.650 | 1.104 | 1.747 | 1.028 | 1.850 | 0.951 | 1.959 | 0.874 | 2.071 | 0.798 | 2.188 | 0.723 | 2.309 | 0.649 | 2.431 |
| 29 | 1.341 | 1.483 | 1.270 | 1.563 | 1.198 | 1.650 | 1.124 | 1.743 | 1.050 | 1.841 | 0.975 | 1.944 | 0.900 | 2.052 | 0.826 | 2.164 | 0.753 | 2.278 | 0.681 | 2.396 |
| 30 | 1.352 | 1.489 | 1.284 | 1.567 | 1.214 | 1.650 | 1.143 | 1.739 | 1.071 | 1.833 | 0.998 | 1.931 | 0.926 | 2.034 | 0.854 | 2.141 | 0.782 | 2.251 | 0.712 | 2.363 |

(continued)

TABLE A1. (Continued)

| n | k=1 $d_L$ | k=1 $d_U$ | k=2 $d_L$ | k=2 $d_U$ | k=3 $d_L$ | k=3 $d_U$ | k=4 $d_L$ | k=4 $d_U$ | k=5 $d_L$ | k=5 $d_U$ | k=6 $d_L$ | k=6 $d_U$ | k=7 $d_L$ | k=7 $d_U$ | k=8 $d_L$ | k=8 $d_U$ | k=9 $d_L$ | k=9 $d_U$ | k=10 $d_L$ | k=10 $d_U$ |
|---|---|---|---|---|---|---|---|---|---|---|---|---|---|---|---|---|---|---|---|---|
| 31 | 1.363 | 1.496 | 1.297 | 1.570 | 1.229 | 1.650 | 1.160 | 1.735 | 1.090 | 1.825 | 1.020 | 1.920 | 0.950 | 2.018 | 0.879 | 2.120 | 0.810 | 2.226 | 0.741 | 2.333 |
| 32 | 1.373 | 1.502 | 1.309 | 1.574 | 1.244 | 1.650 | 1.177 | 1.732 | 1.109 | 1.819 | 1.041 | 1.909 | 0.972 | 2.004 | 0.904 | 2.102 | 0.836 | 2.203 | 0.769 | 2.306 |
| 33 | 1.383 | 1.508 | 1.321 | 1.577 | 1.258 | 1.651 | 1.193 | 1.730 | 1.127 | 1.813 | 1.061 | 1.900 | 0.994 | 1.991 | 0.927 | 2.085 | 0.861 | 2.181 | 0.796 | 2.281 |
| 34 | 1.393 | 1.514 | 1.333 | 1.580 | 1.271 | 1.652 | 1.208 | 1.728 | 1.144 | 1.808 | 1.079 | 1.891 | 1.015 | 1.978 | 0.950 | 2.069 | 0.885 | 2.162 | 0.821 | 2.257 |
| 35 | 1.402 | 1.519 | 1.343 | 1.584 | 1.283 | 1.653 | 1.222 | 1.726 | 1.160 | 1.803 | 1.097 | 1.884 | 1.034 | 1.967 | 0.971 | 2.054 | 0.908 | 2.144 | 0.845 | 2.236 |
| 36 | 1.411 | 1.525 | 1.354 | 1.587 | 1.295 | 1.654 | 1.236 | 1.724 | 1.175 | 1.799 | 1.114 | 1.876 | 1.053 | 1.957 | 0.991 | 2.041 | 0.930 | 2.127 | 0.868 | 2.216 |
| 37 | 1.419 | 1.530 | 1.364 | 1.590 | 1.307 | 1.655 | 1.249 | 1.723 | 1.190 | 1.795 | 1.131 | 1.870 | 1.071 | 1.948 | 1.011 | 2.029 | 0.951 | 2.112 | 0.891 | 2.197 |
| 38 | 1.427 | 1.535 | 1.373 | 1.594 | 1.318 | 1.656 | 1.261 | 1.722 | 1.204 | 1.792 | 1.146 | 1.864 | 1.088 | 1.939 | 1.029 | 2.017 | 0.970 | 2.098 | 0.912 | 2.180 |
| 39 | 1.435 | 1.540 | 1.382 | 1.597 | 1.328 | 1.658 | 1.273 | 1.722 | 1.218 | 1.789 | 1.161 | 1.859 | 1.104 | 1.932 | 1.047 | 2.007 | 0.990 | 2.085 | 0.932 | 2.164 |
| 40 | 1.442 | 1.544 | 1.391 | 1.600 | 1.338 | 1.659 | 1.285 | 1.721 | 1.230 | 1.786 | 1.175 | 1.854 | 1.120 | 1.924 | 1.064 | 1.997 | 1.008 | 2.072 | 0.952 | 2.149 |
| 45 | 1.475 | 1.566 | 1.430 | 1.615 | 1.383 | 1.666 | 1.336 | 1.720 | 1.287 | 1.776 | 1.238 | 1.835 | 1.189 | 1.895 | 1.139 | 1.958 | 1.089 | 2.022 | 1.038 | 2.088 |
| 50 | 1.503 | 1.585 | 1.462 | 1.628 | 1.421 | 1.674 | 1.378 | 1.721 | 1.335 | 1.771 | 1.291 | 1.822 | 1.246 | 1.875 | 1.201 | 1.930 | 1.156 | 1.986 | 1.110 | 2.044 |
| 55 | 1.528 | 1.601 | 1.490 | 1.641 | 1.452 | 1.681 | 1.414 | 1.724 | 1.374 | 1.768 | 1.334 | 1.814 | 1.294 | 1.861 | 1.253 | 1.909 | 1.212 | 1.959 | 1.170 | 2.010 |
| 60 | 1.549 | 1.616 | 1.514 | 1.652 | 1.480 | 1.689 | 1.444 | 1.727 | 1.408 | 1.767 | 1.372 | 1.808 | 1.335 | 1.850 | 1.298 | 1.894 | 1.260 | 1.939 | 1.222 | 1.984 |
| 65 | 1.567 | 1.629 | 1.536 | 1.662 | 1.503 | 1.696 | 1.471 | 1.731 | 1.438 | 1.767 | 1.404 | 1.805 | 1.370 | 1.843 | 1.336 | 1.882 | 1.301 | 1.923 | 1.266 | 1.964 |
| 70 | 1.583 | 1.641 | 1.554 | 1.672 | 1.525 | 1.703 | 1.494 | 1.735 | 1.464 | 1.768 | 1.433 | 1.802 | 1.401 | 1.838 | 1.369 | 1.874 | 1.337 | 1.910 | 1.305 | 1.948 |
| 75 | 1.598 | 1.652 | 1.571 | 1.680 | 1.543 | 1.709 | 1.515 | 1.739 | 1.487 | 1.770 | 1.458 | 1.801 | 1.428 | 1.834 | 1.399 | 1.867 | 1.369 | 1.901 | 1.339 | 1.935 |
| 80 | 1.611 | 1.662 | 1.586 | 1.688 | 1.560 | 1.715 | 1.534 | 1.743 | 1.507 | 1.772 | 1.480 | 1.801 | 1.453 | 1.831 | 1.425 | 1.861 | 1.397 | 1.893 | 1.369 | 1.925 |
| 85 | 1.624 | 1.671 | 1.600 | 1.696 | 1.575 | 1.721 | 1.550 | 1.747 | 1.525 | 1.774 | 1.500 | 1.801 | 1.474 | 1.829 | 1.448 | 1.857 | 1.422 | 1.886 | 1.396 | 1.916 |
| 90 | 1.635 | 1.679 | 1.612 | 1.703 | 1.589 | 1.726 | 1.566 | 1.751 | 1.542 | 1.776 | 1.518 | 1.801 | 1.494 | 1.827 | 1.469 | 1.854 | 1.445 | 1.881 | 1.420 | 1.909 |
| 95 | 1.645 | 1.687 | 1.623 | 1.709 | 1.602 | 1.732 | 1.579 | 1.755 | 1.557 | 1.778 | 1.535 | 1.802 | 1.512 | 1.827 | 1.489 | 1.852 | 1.465 | 1.877 | 1.442 | 1.903 |
| 100 | 1.654 | 1.694 | 1.634 | 1.715 | 1.613 | 1.736 | 1.592 | 1.758 | 1.571 | 1.780 | 1.550 | 1.803 | 1.528 | 1.826 | 1.506 | 1.850 | 1.484 | 1.874 | 1.462 | 1.898 |
| 150 | 1.720 | 1.747 | 1.706 | 1.760 | 1.693 | 1.774 | 1.679 | 1.788 | 1.665 | 1.802 | 1.651 | 1.817 | 1.637 | 1.832 | 1.622 | 1.846 | 1.608 | 1.862 | 1.593 | 1.877 |
| 200 | 1.758 | 1.779 | 1.748 | 1.789 | 1.738 | 1.799 | 1.728 | 1.809 | 1.718 | 1.820 | 1.707 | 1.831 | 1.697 | 1.841 | 1.686 | 1.852 | 1.675 | 1.863 | 1.665 | 1.874 |

| $n$ | $k=11$ $d_L$ | $d_U$ | $k=12$ $d_L$ | $d_U$ | $k=13$ $d_L$ | $d_U$ | $k=14$ $d_L$ | $d_U$ | $k=15$ $d_L$ | $d_U$ | $k=16$ $d_L$ | $d_U$ | $k=17$ $d_L$ | $d_U$ | $k=18$ $d_L$ | $d_U$ | $k=19$ $d_L$ | $d_U$ | $k=20$ $d_L$ | $d_U$ |
|---|---|---|---|---|---|---|---|---|---|---|---|---|---|---|---|---|---|---|---|---|
| 16 | 0.098 | 3.503 | | | | | | | | | | | | | | | | | | |
| 17 | 0.138 | 3.378 | 0.087 | 3.557 | | | | | | | | | | | | | | | | |
| 18 | 0.177 | 3.265 | 0.123 | 3.441 | 0.078 | 3.603 | | | | | | | | | | | | | | |
| 19 | 0.220 | 3.159 | 0.160 | 3.335 | 0.111 | 3.496 | 0.070 | 3.642 | | | | | | | | | | | | |
| 20 | 0.263 | 3.063 | 0.200 | 3.234 | 0.145 | 3.395 | 0.100 | 3.542 | 0.063 | 3.676 | | | | | | | | | | |
| 21 | 0.307 | 2.976 | 0.240 | 3.141 | 0.182 | 3.300 | 0.132 | 3.448 | 0.091 | 3.583 | 0.058 | 3.705 | | | | | | | | |
| 22 | 0.349 | 2.897 | 0.281 | 3.057 | 0.220 | 3.211 | 0.166 | 3.358 | 0.120 | 3.495 | 0.083 | 3.619 | 0.052 | 3.731 | | | | | | |
| 23 | 0.391 | 2.826 | 0.322 | 2.979 | 0.259 | 3.128 | 0.202 | 3.272 | 0.153 | 3.409 | 0.110 | 3.535 | 0.076 | 3.650 | 0.048 | 3.753 | | | | |
| 24 | 0.431 | 2.761 | 0.362 | 2.908 | 0.297 | 3.053 | 0.239 | 3.193 | 0.186 | 3.327 | 0.141 | 3.454 | 0.101 | 3.572 | 0.070 | 3.678 | 0.044 | 3.773 | | |
| 25 | 0.470 | 2.702 | 0.400 | 2.844 | 0.335 | 2.983 | 0.275 | 3.119 | 0.221 | 3.251 | 0.172 | 3.376 | 0.130 | 3.494 | 0.094 | 3.604 | 0.065 | 3.702 | 0.041 | 3.790 |
| 26 | 0.508 | 2.649 | 0.438 | 2.784 | 0.373 | 2.919 | 0.312 | 3.051 | 0.256 | 3.179 | 0.205 | 3.303 | 0.160 | 3.420 | 0.120 | 3.531 | 0.087 | 3.632 | 0.060 | 3.724 |
| 27 | 0.544 | 2.600 | 0.475 | 2.730 | 0.409 | 2.859 | 0.348 | 2.987 | 0.291 | 3.112 | 0.238 | 3.233 | 0.191 | 3.349 | 0.149 | 3.460 | 0.112 | 3.563 | 0.081 | 3.658 |
| 28 | 0.578 | 2.555 | 0.510 | 2.680 | 0.445 | 2.805 | 0.383 | 2.928 | 0.325 | 3.050 | 0.271 | 3.168 | 0.222 | 3.283 | 0.178 | 3.392 | 0.138 | 3.495 | 0.104 | 3.592 |
| 29 | 0.612 | 2.515 | 0.544 | 2.634 | 0.479 | 2.755 | 0.418 | 2.874 | 0.359 | 2.992 | 0.305 | 3.107 | 0.254 | 3.219 | 0.208 | 3.327 | 0.166 | 3.431 | 0.129 | 3.528 |
| 30 | 0.643 | 2.477 | 0.577 | 2.592 | 0.512 | 2.708 | 0.451 | 2.823 | 0.392 | 2.937 | 0.337 | 3.050 | 0.286 | 3.160 | 0.238 | 3.266 | 0.195 | 3.368 | 0.156 | 3.465 |
| 31 | 0.674 | 2.443 | 0.608 | 2.553 | 0.545 | 2.665 | 0.484 | 2.776 | 0.425 | 2.887 | 0.370 | 2.996 | 0.317 | 3.103 | 0.269 | 3.208 | 0.224 | 3.309 | 0.183 | 3.406 |
| 32 | 0.703 | 2.411 | 0.638 | 2.517 | 0.576 | 2.625 | 0.515 | 2.733 | 0.457 | 2.840 | 0.401 | 2.946 | 0.349 | 3.050 | 0.299 | 3.153 | 0.253 | 3.252 | 0.211 | 3.348 |
| 33 | 0.731 | 2.382 | 0.668 | 2.484 | 0.606 | 2.588 | 0.546 | 2.692 | 0.488 | 2.796 | 0.432 | 2.899 | 0.379 | 3.000 | 0.329 | 3.100 | 0.283 | 3.198 | 0.239 | 3.293 |
| 34 | 0.758 | 2.355 | 0.695 | 2.454 | 0.634 | 2.554 | 0.575 | 2.654 | 0.518 | 2.754 | 0.462 | 2.854 | 0.409 | 2.954 | 0.359 | 3.051 | 0.312 | 3.147 | 0.267 | 3.240 |
| 35 | 0.783 | 2.330 | 0.722 | 2.425 | 0.662 | 2.521 | 0.604 | 2.619 | 0.547 | 2.716 | 0.492 | 2.813 | 0.439 | 2.910 | 0.388 | 3.005 | 0.340 | 3.099 | 0.295 | 3.190 |
| 36 | 0.808 | 2.306 | 0.748 | 2.398 | 0.689 | 2.492 | 0.631 | 2.586 | 0.575 | 2.680 | 0.520 | 2.774 | 0.467 | 2.868 | 0.417 | 2.961 | 0.369 | 3.053 | 0.323 | 3.142 |
| 37 | 0.831 | 2.285 | 0.772 | 2.374 | 0.714 | 2.464 | 0.657 | 2.555 | 0.602 | 2.646 | 0.548 | 2.738 | 0.495 | 2.829 | 0.445 | 2.920 | 0.397 | 3.009 | 0.351 | 3.097 |

(continued)

TABLE A1. (Continued)

| n | $k=11$ $d_L$ | $d_U$ | $k=12$ $d_L$ | $d_U$ | $k=13$ $d_L$ | $d_U$ | $k=14$ $d_L$ | $d_U$ | $k=15$ $d_L$ | $d_U$ | $k=16$ $d_L$ | $d_U$ | $k=17$ $d_L$ | $d_U$ | $k=18$ $d_L$ | $d_U$ | $k=19$ $d_L$ | $d_U$ | $k=20$ $d_L$ | $d_U$ |
|---|---|---|---|---|---|---|---|---|---|---|---|---|---|---|---|---|---|---|---|---|
| 38 | 0.854 | 2.265 | 0.796 | 2.351 | 0.739 | 2.438 | 0.683 | 2.526 | 0.628 | 2.614 | 0.575 | 2.703 | 0.522 | 2.792 | 0.472 | 2.880 | 0.424 | 2.968 | 0.378 | 3.054 |
| 39 | 0.875 | 2.246 | 0.819 | 2.329 | 0.763 | 2.413 | 0.707 | 2.499 | 0.653 | 2.585 | 0.600 | 2.671 | 0.549 | 2.757 | 0.499 | 2.843 | 0.451 | 2.929 | 0.404 | 3.013 |
| 40 | 0.896 | 2.228 | 0.840 | 2.309 | 0.785 | 2.391 | 0.731 | 2.473 | 0.678 | 2.557 | 0.626 | 2.641 | 0.575 | 2.724 | 0.525 | 2.808 | 0.477 | 2.829 | 0.430 | 2.974 |
| 45 | 0.988 | 2.156 | 0.938 | 2.225 | 0.887 | 2.296 | 0.838 | 2.367 | 0.788 | 2.439 | 0.740 | 2.512 | 0.692 | 2.586 | 0.644 | 2.659 | 0.598 | 2.733 | 0.553 | 2.807 |
| 50 | 1.064 | 2.103 | 1.019 | 2.163 | 0.973 | 2.225 | 0.927 | 2.287 | 0.882 | 2.350 | 0.836 | 2.414 | 0.792 | 2.479 | 0.747 | 2.544 | 0.703 | 2.610 | 0.660 | 2.675 |
| 55 | 1.129 | 2.062 | 1.087 | 2.116 | 1.045 | 2.170 | 1.003 | 2.225 | 0.961 | 2.281 | 0.919 | 2.338 | 0.877 | 2.396 | 0.836 | 2.454 | 0.795 | 2.512 | 0.754 | 2.571 |
| 60 | 1.184 | 2.031 | 1.145 | 2.079 | 1.106 | 2.127 | 1.068 | 2.177 | 1.029 | 2.227 | 0.990 | 2.278 | 0.951 | 2.330 | 0.913 | 2.382 | 0.874 | 2.434 | 0.836 | 2.487 |
| 65 | 1.231 | 2.006 | 1.195 | 2.049 | 1.160 | 2.093 | 1.124 | 2.138 | 1.088 | 2.183 | 1.052 | 2.229 | 1.016 | 2.276 | 0.980 | 2.323 | 0.944 | 2.371 | 0.908 | 2.419 |
| 70 | 1.272 | 1.987 | 1.239 | 2.026 | 1.206 | 2.066 | 1.172 | 2.106 | 1.139 | 2.148 | 1.105 | 2.189 | 1.072 | 2.232 | 1.038 | 2.275 | 1.005 | 2.318 | 0.971 | 2.362 |
| 75 | 1.308 | 1.970 | 1.277 | 2.006 | 1.247 | 2.043 | 1.215 | 2.080 | 1.184 | 2.118 | 1.153 | 2.156 | 1.121 | 2.195 | 1.090 | 2.235 | 1.058 | 2.275 | 1.027 | 2.315 |
| 80 | 1.340 | 1.957 | 1.311 | 1.991 | 1.283 | 2.024 | 1.253 | 2.059 | 1.224 | 2.093 | 1.195 | 2.129 | 1.165 | 2.165 | 1.136 | 2.201 | 1.106 | 2.238 | 1.076 | 2.275 |
| 85 | 1.369 | 1.946 | 1.342 | 1.977 | 1.315 | 2.009 | 1.287 | 2.040 | 1.260 | 2.073 | 1.232 | 2.105 | 1.205 | 2.139 | 1.177 | 2.172 | 1.149 | 2.206 | 1.121 | 2.241 |
| 90 | 1.395 | 1.937 | 1.369 | 1.966 | 1.344 | 1.995 | 1.318 | 2.025 | 1.292 | 2.055 | 1.266 | 2.085 | 1.240 | 2.116 | 1.213 | 2.148 | 1.187 | 2.179 | 1.160 | 2.211 |
| 95 | 1.418 | 1.930 | 1.394 | 1.956 | 1.370 | 1.984 | 1.345 | 2.012 | 1.321 | 2.040 | 1.296 | 2.068 | 1.271 | 2.097 | 1.247 | 2.126 | 1.222 | 2.156 | 1.197 | 2.186 |
| 100 | 1.439 | 1.923 | 1.416 | 1.948 | 1.393 | 1.974 | 1.371 | 2.000 | 1.347 | 2.026 | 1.324 | 2.053 | 1.301 | 2.080 | 1.277 | 2.108 | 1.253 | 2.135 | 1.229 | 2.164 |
| 150 | 1.579 | 1.892 | 1.564 | 1.908 | 1.550 | 1.924 | 1.535 | 1.940 | 1.519 | 1.956 | 1.504 | 1.972 | 1.489 | 1.989 | 1.474 | 2.006 | 1.458 | 2.023 | 1.443 | 2.040 |
| 200 | 1.654 | 1.885 | 1.643 | 1.896 | 1.632 | 1.908 | 1.621 | 1.919 | 1.610 | 1.931 | 1.599 | 1.943 | 1.588 | 1.955 | 1.576 | 1.967 | 1.565 | 1.979 | 1.554 | 1.991 |

SOURCE: Savin and White (1977, table 3; pp. 1994–1995). Reproduced with permission of the Econometric Society.

TABLE A2. Dickey-Fuller Unit Root *t* Test Statistics for the AR(*p*) Model with No Shift or Deterministic Trend

| N | Probability to the right of critical value | | | | | | | | |
|---|------|------|------|------|------|------|------|------|------|
| | *.01* | *.03* | *.05* | *.10* | *.50* | *.90* | *.95* | *.98* | *.99* |
| 25 | −2.65 | −2.26 | −1.95 | −1.60 | −0.47 | 0.92 | 1.33 | 1.70 | 2.15 |
| 50 | −2.62 | −2.25 | −1.95 | −1.61 | −0.49 | 0.91 | 1.31 | 1.66 | 2.08 |
| 100 | −2.60 | −2.24 | −1.95 | −1.61 | −0.50 | 0.90 | 1.29 | 1.64 | 2.04 |
| 250 | −2.58 | −2.24 | −1.95 | −1.62 | −0.50 | 0.89 | 1.28 | 1.63 | 2.02 |
| 500 | −2.58 | −2.23 | −1.95 | −1.62 | −0.50 | 0.89 | 1.28 | 1.62 | 2.01 |
| ∞ | −2.58 | −2.23 | −1.95 | −1.62 | −0.51 | 0.89 | 1.28 | 1.62 | 2.01 |

SOURCE: Fuller (1996, table 10.A.2, p. 642). Reproduced with permission of John Wiley & Sons.

# NOTES

1. Autocorrelation at lag $k$ is correlation between a time series and its $k$th lag. See chapter 2, Sections 4B and 5A.
2. See chapter 2 for an explanation of independent and identically distributed noise.

## CHAPTER TWO

1. When $k = 0$, the covariance is the variance of $\{X_t\}$.
2. For example, a time series with an autoregressive conditional heteroscedastic (ARCH) process can be white noise but not IID noise (for ARCH($q$) process, see section 4C).
3. For the residual assumptions, see chapter 3.
4. For a unit root process, see chapter 6, section 3B.
5. Examples of standard-alone time series analysis programs are EViews and ITSM. Among multipurpose programs are MATLAB, Minitab, SAS, S+, SPSS, Stata, Statgraphics, and R. EViews is short for Econometric Views. ITSM is short for Interactive Time Series Modeling package. SAS is short for Statistical Analysis System. SPSS is short for Statistical Package for the Social Sciences. Stata is a syllabic abbreviation for Statistics and Data. Free online detailed explanations of how to conduct time series analysis with each of these programs are available (see appendix 1).

Multipurpose programs may require a specific procedure to conduct time series analysis. When we use SPSS, for instance, we should create a date variable before we undertake a time series analysis. SPSS cannot automatically recognize a time series and its periodicity without being informed of a separate date variable. Creating a date variable is a simple task: once we define the appropriate time interval in the Define Dates dialog box, SPSS creates a date variable automatically. In

Minitab, we do not have to create a separate date variable. When analyzing a time series, however, we should let Minitab know what type of time interval we are using by selecting one from among calendar values, clock values, and index values.

6. For the weighted least squares approach, see Neter et al. (1990).

7. In some programs, such as Minitab and ITSM, we need to input the length of periodicity. SPSS determines the length of periodicity by referring to a date variable that we create.

8. Most statistical programs provide the differencing procedure. We usually input a value for the order of differencing, for example, 1 to remove a linear trend and 2 to remove a quadratic trend. Some programs provide a separate seasonal differencing procedure as well. In this case, we need to check whether our program asks us to input the order of differencing or the lag of differencing. If programs ask us to input the lag of differencing (e.g. Minitab and ITSM), we should input 12 for the monthly violent crime rates from 1983 to 1992 (see figures 7 and 11). If programs ask us to input the order of seasonal differencing (e.g. SPSS), we should input 1, not 12. If we input 12 as the order, the first 144 cases will become unavailable.

9. See section 5A for the structure of an autoregressive moving average (ARMA) process.

10. For how to identify systematic patterns of residuals, see section 5.

11. In some AR processes, ACFs may show a fairly slow decay. For example, if ACF at lag 1 is quite strong, that is, if it is 0.9, ACF at 16th lag will be the first ACF that falls inside the 95% confidence limit. In addition, a few occasional ACFs or PACFs, especially at higher lags, may be significantly different from zero by chance alone.

12. Mathematically, the expected number of turning points is 78.7 ($236/3 = 78.7$). However, a turning point which is counted as 0.7 does not actually exist. Statistical programs do not consider this point. When the numbers of observed turning points and/or expected turning points are large, we usually do not have to care about the decimals. We will come to the identical conclusion about the randomness of data, whether or not we include the decimals in calculation. In the case of the detrended and deseasonalized monthly crime rates, we reject the null hypothesis whether we include 0.7 in the calculation ($z = 2.77$) or not ($z = 2.62$),—see table 3. When the numbers of observed turning points and/or expected turning points are small, however, the decimals can make a difference in the turning point test. For example, if the number of observed turning points is 7 and if the expected number of turning points is 14.9 with the standard error equal to 4, we will reject the null hypothesis when we include 0.9 in our calculation ($z = 1.975$) but not when we consider the expected number of turning points that can exist ($z = 1.75$).

13. For an AR(1) process, see section 5.

14. In a few cases, we utilize the presence of unit roots. For example, when two time series variables are cointegrated, we utilize unit roots to avoid the spurious-regression problem (see chapter 6, section 3B). In addition, the idea of unit root can be utilized to test whether a time series tends to be highly persistent (see e.g. Box-Steffensmeier and Smith 1996).

15. If the ACF at lag 1 is $\phi$ and if $\{Z_t\}$ is white noise with a zero mean and variance $\sigma^2$ $(Y_t = \phi Y_{t-1} + Z_t)$, the covariance between $\{Y_t\}$ and $\{Y_{t-k}\}$ is

$$\begin{aligned}
\text{covariance}(Y_t Y_{t-k}) = \text{E}(Y_t Y_{t-k}) &= \text{E}(\phi Y_{t-1} + Z_t)(Y_{t-k}) \\
&= \phi\text{E}(Y_{t-1} Y_{t-k}) + \text{E}(Z_t\, Y_{t-k}) \\
&= \phi\text{E}(Y_{t-1} Y_{t-k}) \\
&= \phi\text{E}(Y_t Y_{t-k+1}) \\
&= \phi\, \text{covariance}(Y_t Y_{t-(k-1)})
\end{aligned}$$

$$\begin{aligned}
\phi\, \text{covariance}(Y_t Y_{t-(k-1)}) = \phi\text{E}(Y_t Y_{t-k+1}) &= \phi[\text{E}(\phi Y_{t-1} + Z_t)(Y_{t-k+1})] \\
&= \phi^2\text{E}(Y_{t-1} Y_{t-k+1}) \\
&= \phi^2\text{E}(Y_t Y_{t-k+2}) \\
&= \phi^2\text{covariance}(Y_t Y_{t-(k-2)})
\end{aligned}$$

$$\cdot$$
$$\cdot$$
$$\cdot$$

$$\phi^{k-1}\, \text{covariance}(Y_t Y_{t-[k-(k-1)]}) = \phi^k\, \text{covariance}(Y_t Y_{t-(k-k)}) = \phi^k\text{variance}(Y_t)$$

Then the ACF at lag $k$ of an AR(1) process is

$$\text{ACF}(k) = \text{covariance}(Y_t Y_{t-k})\, /\, \text{variance}(Y_t) = [\phi^k\text{variance}(Y_t)]\, /\, \text{variance}(Y_t) = \phi^k.$$

16. It is not simple to calculate PACF at lag $k$ in an MA($q$) process. For a pure MA(1) process, $Y_t = Z_t + \theta Z_{t-1}$, the PACF at lag $k$ is

$$\frac{-(-\theta)^k}{(1 + \sum_1^k \theta^{2k})}$$

In a pure MA(1) process, the covariance between $\{Y_t\}$ and $\{Y_{t-1}\}$ is

$$\begin{aligned}
\text{E}(Y_t Y_{t-1}) &= \text{E}(Z_t + \theta Z_{t-1})(Z_{t-1} + \theta Z_{t-2}) \\
&= \text{E}(Z_t Z_{t-1}) + \theta\text{E}(Z_t\, Z_{t-2}) + \theta\text{E}(Z_{t-1})^2 + \theta^2\text{E}(Z_{t-1} Z_{t-2}) \\
&= \theta\text{E}(Z_{t-1})^2 = \theta\sigma^2
\end{aligned}$$

since $\{Z_t\}$ is white noise with zero mean and variance $\sigma^2$, and $\text{E}(Z_t Z_{t-1}) = \text{E}(Z_t Z_{t-2})$ $= \text{E}(Z_{t-1} Z_{t-2}) = 0$.

$$\begin{aligned}
\text{E}(Y_t)^2 &= \text{E}(Z_t + \theta Z_{t-1})^2 \\
&= \text{E}(Z_t)^2 + \theta\text{E}(Z_t\, Z_{t-1}) + \theta\text{E}(Z_{t-1} Z_t) + \theta^2\text{E}(Z_{t-1})^2 \\
&= \text{E}(Z_t)^2 + \theta^2\text{E}(Z_{t-1})^2 \\
&= \sigma^2 + \theta^2\sigma^2
\end{aligned}$$

Then, the PACF at lag 1 is

$$PACF(1) = ACF(1) = E(Y_t Y_{t-1}) / E(Y_t)^2 = \theta\sigma^2 / (\sigma^2 + \theta^2\sigma^2) = \theta / (1 + \theta^2).$$

The covariance between $\{Y_t\}$ and $\{Y_{t-2}\}$ is

$$E(Y_t Y_{t-2}) = E(Z_t + \theta Z_{t-1})( Z_{t-2} + \theta Z_{t-3}) = 0$$

Thus, ACF(2) is zero. Then, the PACF at lag 2 is

$$PACF\ (2) = \frac{ACF(2) - ACF(1)^2}{1 - ACF(1)^2} = \frac{-ACF(1)^2}{1 - ACF(1)^2} = -[\theta/(1 + \theta^2)]^2 / \{1 - [\theta/(1 + \theta^2)]^2\}$$

$$= -\theta^2/(1 + \theta^2 + \theta^4)$$

17. In an MA(2) process, for example,

$$Y_t = Z_t + \theta_1 Z_{t-1} + \theta_2 Z_{t-2},$$
$$Z_{t-1} = Y_{t-1} - \theta_1 Z_{t-2} - \theta_2 Z_{t-3}, \text{ and}$$
$$Z_{t-2} = Y_{t-2} - \theta_1 Z_{t-3} - \theta_2 Z_{t-4}.$$

Thus,

$$Y_t = Z_t + \theta_1(Y_{t-1} - \theta_1 Z_{t-2} - \theta_2 Z_{t-3}) + \theta_2(Y_{t-2} - \theta_1 Z_{t-3} - \theta_2 Z_{t-4})$$
$$= Z_t + \theta_1 Y_{t-1} - \theta_1^2 (Y_{t-2} - \theta_1 Z_{t-3} - \theta_2 Z_{t-4}) - \theta_1\theta_2 Z_{t-3} + \theta_2 Y_{t-2} - \theta_2\theta_1 Z_{t-3}$$
$$\quad + \theta_2^2 Z_{t-4}$$
$$= Z_t + \theta_1 Y_{t-1} - \theta_1^2 Y_{t-2} + \theta_1^3 Z_{t-3} - \theta_1^2\theta_2 Z_{t-4} - \theta_1\theta_2 Z_{t-3} + \theta_2 Y_{t-2} - \theta_2\theta_1 Z_{t-3}$$
$$\quad + \theta_2^2 Z_{t-4}.$$

Since only the lag-1 and lag-2 previous white noise components, $Z_{t-1}$ and $Z_{t-2}$, influence the current observation, $Y_t$,

$$Y_t = Z_t + \theta_1 Y_{t-1} - (\theta_1^2 - \theta_2) Y_{t-2}$$

Thus, in an MA(2) process, lag-1 and lag-2 previous observations influence the current observation through lag-1 and lag-2 previous white noise components. As a result, ACF(1) and ACF(2) will spike, while ACFs whose lags are greater than 2 are not significantly different from zero.

As in the above example, a pure MA($q$) process can be expressed with an AR process of infinite order. Also, a pure AR($p$) process can be expressed with an MA process of infinite order. This characteristic is called the *invertibility* of ARMA($p$, $q$) process. In this respect, an AR model and an MA model are algebraically equivalent. This invertibility, however, does not convey theoretical implication in the social sciences. We select, estimate, and explain an AR model and/or an MA model based on our observations.

18. For further discussion, see Brockwell and Davis (2002), Kohn and Ansley (1986), and Melard (1984).

19. For more selection criteria, see Cromwell, Labys, and Terraza (1994).

20. In this book, we use ARCH($q$)/GARCH($q$, $p$) to avoid the confusion regarding the $p$ and $q$ terms in ARCH/GARCH models and in AR/ARMA models. However, studies may use ARCH($p$)/GARCH($p$, $q$) or ARCH($q$)/GARCH($p$, $q$). We need to check how the $p$ and $q$ terms are defined in studies that employ ARCH/GARCH models.

21. The Lagrange multiplier test is originally a test for an AR($p$) process with the null hypothesis that each of the AR coefficients from lag 1 to lag $p$ is zero and the alternative hypothesis that at least one of them is not zero.

22. For multiple time series analysis, see chapter 6.

23. We are usually unconcerned about preliminary estimates. For this reason, most statistical programs do not report preliminary estimates. Some programs, such as SPSS, report them in their final output of the maximum likelihood estimation procedure.

24. Programs may require us to input three values—$p$, $d$, and $q$—for an ARIMA($p$, $d$, $q$) model, in which $d$ is the order of differencing. As explained above, doing lag-$k$ differencing $d$ times may not just remove trends and seasonal components. It may also influence a systematic pattern of residuals. In addition, we have already estimated and taken out trends and/or seasonal components when we estimate a systematic pattern of residuals. Therefore, we input zero for $d$.

25. For example, SPSS, Minitab, and ITSM produced identical coefficient estimates, except for the MA coefficient of the ARMA(1, 1) model, which was estimated by ITSM as –0.26. This may be because an MA(1) process did not actually exist.

26. The coefficient of the lag-1 term of the squared residuals in the regression of the squared residuals on its lag-1 term is 0.044 ($p = .547$). The coefficients of the lag-1 and lag-2 terms of the squared residuals in the regression of the squared residuals on these two lagged terms are 0.12 ($p = .19$) and 0.02 ($p = .80$), respectively.

CHAPTER FOUR

1. It adds an autoregressive-damping parameter to modify the trend component in Holt's linear trend algorithm (Gardner and McKenzie 1985). If the parameter is less than 1, the algorithm will produce a damped forecast function. If the parameter is greater than 1, the algorithm will produce an exponential forecast function.

2. For example, when observed values are 1, 3, 7, 8, and 10, the MAPE will be 100 when the predicted values are all zero or when predicted values are 2, 6, 14, 16, and 20. The MAPE will be 300 when the predicted values are 4, 12, 28, 32, and 40. The MAPE will be 700 when the predicted values are 8, 24, 56, 64, and 80.

3. In bivariate OLS regression analysis, the MSE is a residual variance. The square root of MSE (RMSE) is a standard error of a slope estimate.

## CHAPTER SIX

1. In this chapter, $\{Y_t\}$ denotes the dependent series and $\{X_t\}$ denotes an independent series, as usual in the OLS regression analysis of cross-sectional data. In chapters 2–5, $\{Y_t\}$ denoted the residuals, namely a deseasonalized and/or detrended series, while $\{X_t\}$ denoted an original series.

2. We can also set up a bidirectional causal relationship between two variables in cross-sectional analysis, if we have the same cross-sectional data accumulated for a long period. However, in this case, we eventually transform cross-sectional analysis into pooled time series analysis.

3. We estimate the $AR(p)$ process after we deseasonalize and/or detrend an independent variable, as explained in chapter 2. Otherwise, trend and/or seasonality will be reflected in the $AR(p)$ process, and the estimated coefficient may be very high.

4. For prewhitening, see section 2.

5. We can mathematically calculate the persistence rate at each lag for the negative AR coefficient of an $AR(1)$ process. However, variables that automatically change the sign of their values from one time point to another are not likely to exist in the social sciences.

6. Calculation of the persistence rate is more complicated when the order of AR process of residuals is greater than 2. We sometimes take more intuitive and less complicated approaches (Pivetta and Reis 2007). For example, we consider only the largest AR coefficient, mostly the lag-1 coefficient. Alternatively, we may use the sum of all significant AR coefficients as an AR coefficient. However, we cannot consider some valuable information of an AR process when we take these approaches, and therefore our estimate of the persistence rate will be less accurate, compared with when we consider all AR coefficients.

7. For the same reason, we do not employ a stepwise procedure that selects independent variables simply according to their statistical significance in multiple regression analysis of cross-sectional data.

8. An omitted third factor, which is positively (negatively) related with an independent variable and negatively (positively) related with the dependent variable, will produce an understated positive or overstated negative coefficient estimate. An omitted third factor, which is positively (negatively) related both with the dependent variable and with an independent variable, will produce an overstated positive or understated negative coefficient estimate. If the relationship between the dependent variable and an independent variable is not real but caused by the intervening role of a third factor, the estimated coefficient of the independent variable will become insignificant once the third factor is controlled out.

9. For a related discussion, see Liu and Hanssens (1982).

10. For how to identify an $ARMA(p, q)$ process, see chapter 2.

11. When we fit an AR(2) model to the detrended consumer price index, the estimated AR coefficients are 1.368 (standard error 0.087) at lag 1 and −0.384 (0.087) at lag 2. In this case, the first AR coefficient is *explosive* ($\phi > 1$), but it is adjusted by the second AR coefficient. The autocorrelation at lag 1 for the estimated AR(1) process, 0.987, is almost identical to that for the estimated AR(2) process, 0.988 (for how to calculate autocorrelations, see chapter 2, section 5A). The plot of the residuals obtained by fitting an AR(2) process to the detrended inflation is identical to figure 48.

12. In the social sciences, the inclusion of lagged dependent variables in time series models has been controversial. Including the lagged dependent variable in a multiple time series model may bias coefficient estimates for independent variables if there is autocorrelation in the residuals (Achen 2000). On the other hand, some support the use of a model with a lagged dependent variable (Beck and Katz 2011; Keele and Kelly 2006).

13. A cause always happens prior to its effect. However, their causal relationship is often considered as instantaneous, especially in the social sciences, simply because the temporal order between them cannot be captured. For example, when a cause happens several days prior to its effect, their relationship will be considered as instantaneous in monthly indices.

14. The instrumental variable is briefly explained in section 1A.

15. This is a frequently cited example for a cointegrating relationship. The case of a drunken woman and her dog may be more appropriate as an example for a cointegrating relationship when they are joined by a leash with a retractable cord than when they freely wander. The retractable cord will offer them a limited degree of freedom on the leash.

16. The constant, $a$, will not make a difference in the estimation of $b_3$.

17. $-1.2 \times -0.2 = 0.24$,
$(-1.2 + 0.24) \times -0.2 = 0.192$,
$(-0.96 + 0.192) \times -0.2 = 0.154$,
$(-0.768 + 0.154) \times -0.2 = 0.123$,
$(-0.614 + 0.123) \times -0.2 = 0.098$, etc.

18. For deseasonlization, see chapter 2, section 3A.

19. When testing whether two variables are cointegrated, we refer to Dickey-Fuller test statistics calculated without a constant and a deterministic trend (see table A2, appendix 4, page 209).

The Dickey-Fuller test that includes a constant and a trend is:

$$\nabla_1 Y_t = b_0 + b_1 T + (\phi - 1)Y_{t-1} + e_t$$

where $T$ is time and $e_t$ is white noise.

With this test, we test whether $Y_t$ is stationary around a deterministic trend that is a nonrandom function of time and therefore predictable. A deterministic trend in the Dickey-Fuller test means that $Y_t$, could be growing steadily over time, which violates the idea of cointegration.

20. Mathematical discussion of unit-root processes and near or fractional unit-root processes is beyond the scope of this book. For detailed discussion about near or fractional integration, see Box-Steffensmeier and Smith (1998); De Boef and Granato (1997, 1999); Diebold and Rudebusch (1989); Granger (1980); Granger and Ding (1996); Granger and Joyeux (1980); Joyeux (2010).

21. For a related discussion, see section 3C below.

22. This error-correction model is different from a model that we can employ to solve the autocorrelated-residuals problem, such as a model with a lagged dependent variable and lagged independent variables on the right-hand side and the Cochrane-Orcutt model (see section 4).

23. In the social sciences, we often include variables measured at the interval or ordinal level in our models, such as OLS regression models that require data to be measured at the ratio level.

24. As shown in chapter 2, the monthly violent crime rates contain an AR(2) process. Therefore, we may try the lag-1 and lag-2 previous terms of the dependent variable. When we prewhiten variables in our model, however, we do not need to consider the AR process, and it will be enough to include the lag-1 previous dependent variable to control the continuity of the dependent variable.

25. As explained above, the Granger causality was originally applied to stationary time series. However, trends are often of major concern in social science studies. In fact, when we detrended the deseasonalized monthly violent crime rates and detrended two independent variables, only the lagged dependent variable was statistically significant. We need to take this result with reservation because we eliminated the most important part of the dependent variable that we expect the independent variables to explain and the most important part of two independent variables that we expect to explain the dependent variable.

26. According to figure 53, an AR(1) process may also represent the stochastic process of the residuals. In this case, the AR coefficient is 0.706 (standard error 0.066).

27. The coefficient of inflation in the Bewley transformation regression is 0.496 ($\approx 0.474$).

28. Systematic patterns in residuals can be represented by an AR($p$) process and/or an MA($q$) process. In the social sciences, the AR($p$) process is more frequently employed than the MA($q$) process simply because it is easier to estimate the AR($p$) process (Beck 1991).

29. This Durbin $h$ statistic was calculated with the Durbin-Watson $d$ statistic, 1.99, and the variance of the lagged dependent variable, 0.073:

$$h = \left(1 - \frac{1.99}{2}\right)\sqrt{\frac{120}{1-120\times(.073^2)}} = .005\times18.2442 = 0.0912$$

30. It is 0.629 (standard error 0.072) when the monthly violent crime rates are not deseasonalized.

31. The adjusted multiple time series model with the estimated coefficient of the AR(1) process, 0.629, is:

$\hat{Y}_t = -5.46 + 0.201(\text{unemployment rate}) + 0.486(\text{inflation}) + 0.629Et{-}1 + Zt$

where $Z_t = \text{WN}(0, 9.66)$.

32. Standard error for unemployment is 0.709. The constant in the regression with adjusted variables was –0.988, with standard error 1.931. The adjusted constant is –3.41 (0.988/0.29), with adjusted standard error 6.66 (1.931/0.29).

33. The Breusch-Godfrey statistic for an AR(1) process in $\{E_t\}$:

$$(120 - 1) \times 0.397 = 47.243.$$

The Breusch-Godfrey statistic for an AR(1) process in $\{Z_t\}$:

$$(120 - 1) \times 0.003 = 0.357.$$

CHAPTER SEVEN

1. These states are Washington (1993); California, Colorado, Connecticut, Georgia, Indiana, Kansas, Louisiana, Maryland, New Mexico, North Carolina, Tennessee, Virginia, and Wisconsin (1994); and Arkansas, Florida, Montana, Nevada, New Jersey, North Dakota, Pennsylvania, South Carolina, Utah, and Vermont (1995). Texas has had its getting-tough-on-crime law since 1974 and revised its penal code in 1994. Two additional states enacted the TSO later: Arizona in 2006 and Massachusetts in 2012.

2. The result will be identical when we include the first 117 cases from January 1985 to September 1994 in the modeling procedure and then compare the monthly violent crime rates from October 1994 to December 2004 with the counterfactuals—the values predicted by the estimated model.

3. The 109th to 117th observations will attenuate the slope of the before-the-intervention crime rates. If we estimate the before-the-intervention univariate time series model without these nine observations, the slope increases from 0.151 to 0.158. The intercept decreases from 47.633 to 47.358. As a result, the forecasted crime rate in September 1994 increases a little, to 65.84 crimes per 100,000 inhabitants, and the drop in level after the intervention is about 9.68 points.

4. The predicted crime rate in December 2004 is 47.633 + (0.151 × 240) = 83.873 crimes per 100,000 inhabitants. It is 85.278 when the before-the-intervention model is estimated without the 109th to 117th observations. The observed crime rate in December 2004 is 36.6 crimes per 100,000 inhabitants.

# REFERENCES

Abramson, P. R., and C. W. Ostrom Jr. 1991. "Macropartisanship: An Empirical Reassessment." *American Political Science Review* 85:181–192.

Abu Sada, M. S. 1998. "Party Identification and Political Attitudes in an Emerging Democracy: A Summary." *American Journal of Political Science* 42:712–715.

Achen, C. H. 1982. *Interpreting and Using Regression.* Sage University Paper Series on Quantitative Applications in the Social Sciences, no. 29. Newbury Park, CA: Sage.

———. 2000. "Why Lagged Dependent Variables Can Suppress the Explanatory Power of Other Independent Variables." Presented at the Annual Meeting of the Political Methodology Section of the American Political Science Association, Los Angeles, July 20–22.

Akaike, H. 1969. "Fitting Autoregressions for Predictions." *Annals of the Institute of Statistical Mathematics* 21:243–247.

———. 1973. "Information Theory and an Extension of the Maximum Likelihood Principle." In *The Second International Symposium on Information Theory*, edited by B. N. Petrov and F. Csaki, 267–281. Budapest: Akademial kiado.

———. 1979. "A Bayesian Extension of the Minimum AIC Procedure." *Biometrika* 66:237–242.

Aldrich, J. H., and F. D. Nelson. 1984. *Linear Probability, Logit, and Probit Models.* Sage University Paper Series on Quantitative Applications in the Social Sciences, no. 45 Newbury Park, CA: Sage.

Armstrong, J. S. 1985. *Long-range Forecasting: From Crystal Ball to Computer.* New York: Wiley.

Austin, J., J. Clark, P. Hardyman, and D. A. Henry. 1999. "The Impact of Three Strikes and You're Out." *Punishment and Society* 1(2):131–162.

Bartels, B. L., J. M. Box-Steffensmeier, C. D. Smidt, and R. M. Smith. 2010. "The Dynamic Properties of Individual-Level Party Identification in the United States." *Electoral Studies* 30:210–222.

Beck, N. 1991. "Comparing Dynamic Specifications: The Case of Presidential Approval." *Political Analysis* 3:51–87.

———. 1992. "The Methodology of Co-integration." *Political Analysis* 4:237–248.

Beck, N., and J. N. Katz. 2011. "Modeling Dynamics in Time-Series-Cross-Section Political Economy Data." *Annual Review of Political Science* 14:331–352.

Berry, W. D. 1993. *Understanding Regression Assumptions*. Sage University Paper Series on Quantitative Applications in the Social Sciences, no. 92. Newbury Park, CA: Sage.

Berry, W. D., and S. Feldman. 1985. *Multiple Regression in Practice*. Sage University Paper Series on Quantitative Applications in the Social Sciences, no. 50. Newbury Park, CA: Sage.

Bewley, R. A. 1979. "The Direct Estimation of the Equilibrium Response in a Linear Model." *Economic Letters* 3:357–361.

Blalock, H. M., Jr. 1979. *Social Statistics*. Revised 2nd ed. New York: McGraw-Hill.

Bollerslev, T. 1986. "Generalized Autoregressive Conditional Heteroskedasticity." *Journal of Econometrics* 31:307–327.

Box, G. E. P., and D. R. Cox. 1964. "An Analysis of Transformations." *Journal of the Royal Statistical Society* 26:211–243.

Box, G. E. P., and G. M. Jenkins. 1976. *Time Series Analysis: Forecasting and Control*. Rev. ed. San Francisco, CA: Holden-Day.

Box, G. E. P., and D. Pierce. 1970. "Distribution of Autocorrelations in Autoregressive Moving Average Time Series Models." *Journal of the American Statistical Association* 65:1509–1526.

Box, G. E. P., and G. C. Tiao. 1975. "Intervention Analysis with Applications to Economic and Environmental Problems." *Journal of the American Statistical Association* 70:70–79.

Box-Steffensmeier, J. M., and R. M. Smith. 1996. "The Dynamics of Aggregate Partisanship." *American Political Science Review* 90:567–580.

———. 1998. "Investigating Political Dynamics Using Fractional Integration Methods." *American Journal of Political Science* 42:661–689.

Brandt, P. T., and J. T. Williams. 2001. "A Linear Poisson Autoregressive Model: The Poisson AR(p) Model." *Political Analysis* 9:164–184.

Breusch, T. S. 1978. "Testing for Autocorrelation in Dynamic Linear Models." *Australian Economic Papers* 17:334–355.

Brockwell, P. J., and R. A. Davis. 2002. *Introduction to Time Series and Forecasting*, 2nd ed. New York: Springer.

Brown, R. G. 1963. *Smoothing, Forecasting, and Prediction*. Englewood Cliffs: Prentice Hall.

Bureau of Justice Statistics, Department of Justice. 1993. *Sourcebook of Criminal Justice Statistics, 1992.*

Burnham, K. P., and D. R. Anderson. 2004. "Multimodel Inference: Understanding AIC and BIC in Model Selection." *Sociological Methods and Research* 33:261–304.

Campbell, A., P. E. Converse, W. E. Miller, and D. E. Stokes. 1960. *The American Voter*. Chicago: Mid-way Reprint.

Chatfield, C. 2004. *The Analysis of Time Series: An Introduction*. 6th ed. Washington, DC: Chapman & Hall/CRC.

Chen, E. Y. 2008. "Impacts of Three Strikes and You're Out on Crime Trends in California and throughout the United States." *Journal of Contemporary Criminal Justice* 24:345–370.

Clarke, H. D., and M. C. Stewart. 1995. "Economic Evaluations, Prime Ministerial Approval and Governing Party Support: Rival Models Reconsidered." *British Journal of Political Science* 25:145–170.

———. 1996. "Economists and Electorates: The Subjective Economy of Governing Party Support in Canada." *European Journal of Political Research* 29:191–214.

Clarke, H.D., M. C. Stewart, and P. Whitely. 1997. "Tory Trends: Party Identification and the Dynamics of Conservative Support Since 1992." *British Journal of Political Science* 27:299–319.

Cochrane, D., and G. H. Orcutt. 1949. "Application of Least Squares Regression to Relationships Containing Auto-Correlated Error Terms." *Journal of the American Statistical Association* 44:32–61.

Colton, T. 2000. *Transitional Citizens: Voters and What Influences Them in the New Russia*. Cambridge, MA: Harvard University Press.

Cook, T. D., and D. T. Campbell. 1979. *Quasi-Experimentation: Design & Analysis Issues for Field Settings*. Boston: Houghton Mifflin Company.

Cox, G. W., and M. D. McCubbins. 1991. "On the Decline of Party Voting in Congress." *Legislative Studies Quarterly*, 16:547–570.

Cromwell, J. B., W. C. Labys, and M. Terraza. 1994. *Univariate Tests for Time Series Models*. Sage University Paper Series on Quantitative Applications in the Social Sciences, no. 99. Newbury Park, CA: Sage.

De Boef, S., and J. Granato. 1997. "Near-Integrated Data and the Analysis of Political Relationships." *American Journal of Political Science* 41:619–640.

———. 1999. "Testing for Co-integrating Relationships with Near-Integrated Data." *Political Analysis* 8:99–117.

De Boef, S., and L. Keele. 2008. "Taking Time Seriously." *American Journal of Political Science* 52:184–200.

Dickey, D. A., and W. A. Fuller. 1979. "Distribution of the Estimators for Autoregressive Time Series with a Unit Root." *Journal of the American Statistical Association* 74:427–431.

Diebold, F. X., and G. D. Rudebusch. 1989. "Long Memory and Persistence in Aggregate Output." *Journal of Monetary Economics* 24:189–209.

Durbin, J. 1970. "Testing for Serial Correlation in Least-Squares Regressions When Some of the Regressors Are Lagged Dependent Variables." *Econometrica* 38:410–421.

Durbin, J., and G. S. Watson. 1951. "Testing for Serial Correlation in Least Squares Regression II." *Biometrika* 38:159–178.

Durr, R. H. 1993. "What Moves Policy Sentiment?" *American Political Science Review* 87:158–170.

Engle, R. F. 1982. "Autoregressive Conditional Heteroskedasticity with Estimates of the Variance of United Kingdom Inflation." *Econometrica* 50:987–1007.

Engle, R. F., and C. W. J. Granger. 1987. "Co-integration and Error Correction: Representation, Estimation and Testing." *Econometrica* 55:251–276.

Erdem, O., E. Ceyhan, and Y. Varli. 2014. "A New Correlation Coefficient for Bivariate Time-Series Data." Available at SSRN (http://ssrn.com/abstract=2390065) or http://dx.doi.org/10.2139/ssrn.2390065.

Ekland-Olson, S., and W. R. Kelly. 1992. "Crime and Incarceration: Some Comparative Findings from the 1980s." *Crime & Delinquency* 38:392–416.

Eubank, R. B., and D. J. Gow. 1983. "The Pro-Incumbent Bias in the 1978 and 1980 National Election Studies." *American Journal of Political Science* 27:122–139.

Fiorina, M. P. 1982. *Retrospective Voting in American National Elections.* New Haven, CT: Yale University Press.

Forehand, G. 1982. *New Directions for Program Evaluation: Applications of Time Series Analysis to Evaluation.* San Francisco: Jossey-Bass.

Fox, J. 1991. *Regression Diagnostics.* Sage University Paper Series on Quantitative Applications in the Social Sciences, no. 79. Newbury Park, NJ: Sage.

Fuller, W. A. 1996. *Introduction to Statistical Time Series.* 2nd ed. New York: John Wiley.

Gardner, E. S., Jr., and E. McKenzie. 1985. "Forecasting Trends in Time Series." *Management Science* 31:1237–1246.

Godfrey, L. G. 1978. "Testing against General Autoregressive and Moving Average Error Models When the Regressors Include Lagged Dependent Variables." *Econometrica* 46:1293–1302.

Goldfield, A., and R. Quandt. 1965. "Some Tests for Heteroscedasticity." *Journal of American Statistical Association* 60:539–547.

Granger, C. W. J. 1969. "Investigating Causal Relations by Econometric Models and Cross-spectral Methods." *Econometrica* 37:424–438.

———. 1980. "Long Memory Relationships and the Aggregation of Dynamic Models." *Journal of Econometrics* 14:227–238.

Granger, C. W. J., and Z. Ding. 1996. "Varieties of Long Memory Models." *Journal of Econometrics* 73:61–77.

Granger, C. W. J., and R. Joyeux. 1980. "An Introduction to Long-Memory Time Series Models and Fractional Differencing." *Journal of Time Series Analysis* 1:15–29.

Granger, C. W. J., and P. Newbold. 1974. "Spurious Regressions in Econometrics." *Journal of Econometrics* 2:111–120.

Green, D. P., A. S. Gerber, and S. L. De Boef. 1999. "Tracking Opinion over Time: A Method for Reducing Sampling Error." *Public Opinion Quarterly* 63:178–192.

Green, D. P., B. Palmquist, and E. Schickler. 2001. "Partisan Stability: Evidence from Aggregate Data." In *Controversies in Voting Behavior*, 4th ed., edited by Richard G. Niemi and Herbert F. Weisberg, 356–363. Washington, DC: CQ Press.

Greenwood, P. W. 1994. *Three Strikes and You're Out: Estimated Benefits and Costs of California's New Mandatory-Sentencing Law.* Santa Monica: RAND.

Gronke, P. 1992. "Overreporting the Vote in the 1988 Senate Election Study: A Response to Wright." *Legislative Studies Quarterly* 17:113–129.

Gronke, P., and J. Brehm. 2002. "History, Heterogeneity, and Presidential Approval: A Modified ARCH Approach." *Electoral Studies* 21:425–452.

Honaker, J., and G. King. 2010. "What to Do about Missing Values in Time-Series Cross- Section Data." *American Journal of Political Science* 54:561–581.

Hurvich, C. M., and C. L. Tsai. 1989. "Regression and Time Series Model Selection in Small Samples." *Biometrika* 76:297–307.

Johansen, S. 1988. "Statistical Analysis of Co-integration Vectors." *Journal of Economic Dynamics and Control* 12:231–254.

Johnston, J. 1984. *Econometric Methods.* 2nd ed. New York: Macmillan.

Joyeux, R. 2010. "Long Memory Processes: A Joint Paper with Clive Granger." *Journal of Financial Econometrics* 8:184–186.

Keele, L., and N. J. Kelly. 2006. "Dynamic Models for Dynamic Theories: The Ins and Outs of Lagged Dependent Variables." *Political Analysis* 14:186–205.

Kellough, J. E. 1989. "The 1978 Civil Service Reform and Federal Equal Employment Opportunity." *American Review of Public Administration* 19:313–324.

Kernell, S. 1978. "Explaining Presidential Popularity." *American Political Science Review* 72:506–522.

Key, V. O. 1966. *The Responsible Electorate.* Cambridge, MA: Harvard University Press.

King, G., J. Honaker, A. Joseph, and K. Scheve. 2001. "Analyzing Incomplete Political Science Data: An Alternative Algorithm for Multiple Imputation." *American Political Science Review* 95:49–69.

King, G., R. Koehane, and S. Verba. 1994. *Designing Social Inquiry: Scientific Inference in Qualitative Research.* Princeton, NJ: Princeton University press.

Kitschelt, H. 1995. "Formation of Party Cleavages in Post-Communist Democracies: Theoretical Propositions." *Party Politics* 1:447–472.

Kohn, R., and C. Ansley. 1986. "Estimation, Prediction, and Interpolation for ARIMA Models with Missing Data." *Journal of the American Statistical Association* 81:751–761.

Kovandzic, T.V., J. J. Sloan III, and L. M. Vieraitis. 2004. "Striking Out as Crime Reduction Policy: The Impact of Three Strikes Laws on Crime Rates in U.S. Cities." *Justice Quarterly* 21:207–239.

Kramer, G. H. 1971. "Short-term Fluctuations in U.S. Voting Behavior, 1896–1964." *American Political Science Review* 65:131–143.

Lewis-Beck, M. 1980. *Applied Regression: An Introduction.* Sage University Paper Series on Quantitative Applications in the Social Sciences, no. 22. Newbury Park, CA: Sage.

———. 1988. "Economics and the American Voter: Past, Present, Future." *Political Behavior* 10:5–21.

Lewis-Beck, M. and J. R. Alford. 1980. "Can Government Regulate Safety? The Coal Mine Example." *American Political Science Review* 74: 745–756.

Little, R. J. A., and D. B. Rubin. 2002. *Statistical Analysis with Missing Data.* 2nd ed. New York: Wiley & Sons.

Liu, L. M., and D. H. Hanssens. 1982. "Identification of Multiple-Input Transfer Function Models." *Communications in Statistics—Theory and Methods* 11:297–314.

Lockerbie, B. 1989. "Change in Party Identification: The Role of Prospective Economic Evaluations." *American Politics Quarterly* 17:291–311.

———. 1992. "Prospective Voting in Presidential Elections: 1956–1988." *American Politics Quarterly* 20:308–325.

———. 2002. "Party Identification: Constancy and Change." *American Politics Research* 30:384–405.

———. 2004. "A Look to the Future: Forecasting the 2004 Presidential Election." *PS: Political Science and Politics* 37:741–45.

———. 2008. "Election Forecasting: The Future of the Presidency and the House." *PS: Political Science and Politics* 41:713–771.

Ljung, G. M., and G. E. P. Box. 1978. "On a Measure of Lack of Fit in Time Series Models." *Biometrika* 65:297–303.

MacKinnon, J. G. 1996. "Numerical Distribution Functions for Unit Root and Co-integration Tests." *Journal of Applied Econometrics* 11:601–618.

Mackuen, M. B., R. S. Erikson, and J. A. Stimson. 1989. "Macropartisanship." *American Political Science Review* 83:1125–1142.

Maestas, C., and R. Preuhs. 2000. "Modeling Volatility in Political Time Series." *Electoral Studies* 19:95–110.

Maguire, K., A. L. Pastore, and T. J. Flanagan (eds.). 1993. *Sourcebook of Criminal Justice Statistics 1992*. Washington, DC: U.S. Department of Justice, Bureau of Justice Statistics.

Mainwaring, S. 1999. *Rethinking Party Systems in the Third Wave of Democratization: The Case of Brazil*. Stanford, CA: Stanford University Press.

Mair, P. 1996. "What Is Different about Post-communist Systems?" *Studies in Public Policy* no. 259. Glasgow: University of Strathclyde, Centre for the Study of Public Policy.

Makridakis, S., S. C. Wheelwright, and V. McGee. 1983. *Forecasting: Methods and Applications*. 2nd ed. New York: Wiley.

Markus, G. B. 1988. "The Impact of Personal and National Economic Conditions on the Presidential Vote: A Pooled Cross-sectional Analysis." *American Journal of Political Science* 32:137–154.

Markus, G. B., and P. E. Converse. 1979. "A Dynamic Simultaneous Equation Model of Vote Choice." *American Political Science Review* 73:1055–1070.

McCleary, R., and J. E. Riggs. 1982. "The 1975 Australian Family Law Act: A Model for Assessing Legal Impacts." In *New Directions for Program Evaluation: Applications of Time Series Analysis to Evaluation*, edited by G. Forehand, 7–18. San Francisco: Jossey-Bass.

McDowall, D., R. McCleary, E. E. Meidinger, and R. A. Hay Jr. 1980. *Interrupted Time Series Analysis*. Sage University Paper Series on Quantitative Applications in the Social Sciences, no. 21. Newbury Park, CA: Sage.

McLeod, A. I., and W. K. Li. 1983. "Diagnostic Checking ARMA Time Series Models Using Squared Residual Autocorrelations." *Journal of Time Series Analysis* 4:269–273.

Melard, G. 1984. "A Fast Algorithm for the Exact Likelihood of Autoregressive Moving Average Models." *Applied Statistics* 33:104–114.

Menard, S. 1995. *Applied Logistic Regression Analysis*. Sage University Paper Series on Quantitative Applications in the Social Sciences, no. 106. Newbury Park, CA: Sage.

Miller, A. H., and T. F. Klobucar. 2000. "The Development of Party Identification in Post-Soviet Societies." *American Journal of Political Science* 44:667–686.

Mohr, L. B. 1992. *Impact Analysis for Program Evaluation*. Newbury Park, CA: Sage.

Mueller, J. E. 1970. "Presidential Popularity from Truman to Johnson." *American Political Science Review* 64:18–34.

———. 1973. *War, Presidents, and Public Opinion*. New York: Wiley.

Murray, M. P. 1994. "A Drunk and Her Dog: An Illustration of Co-integration and Error Correction." *American Statistician* 48:37–39.

Neter, J., W. Wasserman, and M. H. Kutner. 1990. *Applied Linear Statistical Models: Regression, Analysis of Variance, and Experimental Designs*. 3rd ed. Boston: Irwin.

Norpoth, H. 1995. "Is Clinton Doomed? An Early Forecast for 1996." *PS: Political Science & Politics* (June): 201–207.

Norpoth, H., and T. Yantek. 1983. "Macroeconomic Conditions and Fluctuations of Presidential Popularity: The Question of Lagged Effects." *American Journal of Political Science* 27:785–807.

Ostrom, C., and R. Smith. 1992. "Error Correction, Attitude Persistence, and Executive Rewards and Punishments: A Behavioral Theory of Presidential Approval." *Political Analysis* 4:127–183.

Phillips, A. W. 1958. "The Relation between Unemployment and the Rate of Change of Money Wage Rates in the United Kingdom, 1861–1957." *Economica* 25:283–299.

Pivetta, F., and R. Reis. 2007. "The Persistence of Inflation in the United States." *Journal of Economic Dynamics and Control* 31:1326–1358.

Prais, S. J., and C. B. Winsten. 1954. *Trend Estimators and Serial Correlation*. Chicago, IL: Cowles Commission.

Rajmaira, S., and M. D. Ward. 1990. "Evolving Foreign Policy Norms: Reciprocity in the Superpower Triad." *International Studies Quarterly* 34:457–475.

Ramirez, J. R., and W. D. Crano. 2003. "Deterrence and Incapacitation: An Interrupted Time- Series Analysis of California's Three Strikes Law." *Journal of Applied Social Psychology* 33:110–145.

Razali, N. M., and Y. B. Wah. 2011. "Power Comparisons of Shapiro-Wilk, Kolmogorov- Smirnov, Lilliefors and Anderson-Darling Tests." *Journal of Statistical Modeling and Analytics* 2:21–33.

Rosenstone, S. J. 1983. *Forecasting Presidential Elections*. New Haven, CT: Yale University Press.

Rossi, P. H., and H. E. Freeman. 1993. *Evaluation: A Systematic Approach.* 5th edition. Newbury Park, CA: Sage.

Savin, N. E., and K. J. White. 1977. "The Durbin-Watson Test for Serial Correlation with Extreme Sample Sizes or Many Regressors." *Econometrica* 45:1989–1996.

Sayrs, L.W. 1989. *Pooled Time Series Analysis.* Sage University Paper Series on Quantitative Applications in the Social Sciences, no. 70. Newbury Park, CA: Sage.

Schiraldi, V., J. Colburn, and E. Lotke. 2004. *3 Strikes & You're Out: An Examination of the Impact of Strikes Laws 10 years after Their Enactment.* Washington, DC: Justice Policy Institute.

Schwarz, G. 1978. "Estimating the Dimension of a Model." *Annals of Statistics* 6:461–464.

Seier, E. 2002. "Comparison of Tests for Univariate Normality." *InterStat Statistical Journal* 1:1–17.

Shibata, R. 1976. "Selection of the Order of an Autoregressive Model by Akaike's Information Criterion." *Biometrika* 63:117–126.

Shin, Y. S. 2012. "Is Party Identification Responsible in Korea? An Explanation of Unstable Party Identification in Terms of Voters' Evaluations." *Korea Observer* 43:233–252.

Stanley, H. 1988. "Southern Partisan Changes: Dealignment, Realignment or Both?" *Journal of Politics* 50:64–88.

Stolzenberg, L., and S. J. D'Alessio. 1997. "'Three Strikes and You're Out': The Impact of California's New Mandatory Sentencing Law on Serious Crime Rates." *Crime and Delinquency* 43:457–469.

Tomsa, D., and A. Ufen. 2013. *Party Politics in Southeast Asia: Clientelism and Electoral Competition in Indonesia, Thailand and the Philippines.* New York: Routledge.

Weisberg, H. F., and C. C. Smith. "The Influence of the Economy on Party Identification in the Reagan Years." *Journal of Politics* 53:1077–1092.

Williams, J. T. 1992. "What Goes Around Comes Around: Unit Root Tests and Co-integration." *Political Analysis* 4:229–235.

Wlezien, C. 1996. "Dynamics of Representation: The Case of U.S. Spending on Defense." *British Journal of Political Science* 26:81–103.

Worrall, J. L. 2004. "The Effect of Three-Strikes Legislation on Serious Crime in California." *Journal of Criminal Justice* 32: 283–296.

Wright, G. C. 1990. "Misreports of Vote Choice in the 1988 NES Senate Election Study." *Legislative Studies Quarterly* 15:543–564.

———. 1992. "Reported versus Actual Vote: There Is a Difference and It Matters." *Legislative Studies Quarterly* 17:131–142.

Zimring, T. E., S. Karnin, and G. Hawkins. 1999. *Crime and Punishment in California: The Impact of Three Strikes and You're Out.* Berkeley, CA: Institute of Governmental Studies.

# INDEX

Maximum likelihood estimates, 60, 61, 62
McLeod-Li test, 40, 41
Mean absolute error (MAE), 95, 98, 99
Mean absolute percentage error (MAPE),
95, 96, 97, 99
Mean squared error (MSE), 97, 99
Model fit, *see* forecast accuracy
Model selection criteria, 62, 63, 64, 65, 66
Moving average process, 30, 31, 52; correlo-
gram of, 36, 37, 38, 55, 58, 59
Moving average smoothing, 107, 108; *see*
exponential smoothing
Multiple time-series analysis, 115, 116, 117,
118, 119; *see* lagged dependent variable
and lagged independent variable
Multivariate time-series analysis, 115

No autocorrelation assumption; univariate
time series analysis, 79, 80, 81; multiple
time series analysis, 165, 166, 167, 168,
169, 170
Non-constant variance, 16, 19, 20, 21,
22, 23
Normal Q-Q plot, 84
Normal P-P plot, 84
Normality assumption, 81, 82, 83, 84

Optimization process, *see* maximum likeli-
hood estimates
Order of differencing, 32, 33
Outliers, 16, 17, 84
Over-differencing, 35

PACF, *see* partial autocorrelation
Partial autocorrelation, 36, 54
Periodic component, *see* seasonality
Persistence rate, 118, 119, 216n6
Polynomial trend, 28
Portmanteau test, 40
Prais-Winsten procedure, 169, 170
Preliminary definition, 10, 11, 12, 13
Preliminary visual analysis, 15, 16, 17,
18, 19
Prewhitening, 80, 81, 127, 128, 129, 130, 131,
132, 133, 134, 135, 146, 156; *see* equal
footing approach and systems
approach
Prior moving average smoothing, 107, 108

Random walks, 146
Randomness tests, 38, 39, 40, 41, 42, 43,
44, 45, 46, 47, 48, 49
Rank test, 43, 44
Residual assumptions; univariate time
series analysis, 79, 80, 81, 82, 83, 84;
multiple time series analysis, 164, 165,
166, 167, 168, 169, 170
Residual diagnostics, *see* residual
assumptions
Residual variance, 62, 63
Residuals, 11, 12; *see* concordance of system-
atic patterns and discordance of system-
atic patterns
RMSE, *see* square root of MSE
Runs test, 45, 46

Sample ACF and PACF correlograms, 19,
*see* autoregressive process, autoregressive
moving average process, and moving
average process
SBC, *see* Schwarz's Bayesian criterion
Schwarz's Bayesian criterion (SBC), 65
Seasonal dummy variable scheme (seasonal
indicator), 134, 135
Seasonality, 16, 19, 24, 25, 26, 131, 132, 133,
134, 135
Selection criteria, 62, 63, 64, 65, 66
Serial correlation, *see* autocorrelation
Short-memory process, 154
Short-term influence, 147
Simple (seasonal) exponential smoothing
algorithm, 92, 93
Skewness, 82
Smoothing, 106; *see* moving average
smoothing and exponential smoothing
Smoothing constant, 109, 110
Smoothing weight, 109, 110
Square root of MSE (RMSE), 97, 98, 99,
100
Stationarity, 12, 13; *see* difference stationary,
strictly stationary, trend stationary,
weakly stationary
Stem-and-leaf plot, 83, 84
Stochastic trend, 11
Strictly stationary, 12
Symmetric mean absolute percentage error
(SMAPE), 96, 97, 98, 99

9 780520 293168